Rocks, Gems, and Minerals of the Rocky Mountains

Garret Romaine

FALCONGUIDES

GUILFORD, CONNECTICUT
HELENA, MONTANA

AN IMPRINT OF ROWMAN & LITTLEFIELD

FALCONGUIDES®

Copyright © 2014 Rowman & Littlefield

ALL RIGHTS RESERVED. No part of this book may be reproduced or transmitted in any form by any means, electronic or mechanical, including photocopying and recording, or by any information storage and retrieval system, except as may be expressly permitted in writing from the publisher.

FalconGuides is an imprint of Rowman & Littlefield.
Falcon, FalconGuides, and Outfit Your Mind are registered trademarks of Rowman & Littlefield.

Photos by Garret Romaine unless noted otherwise.

Distributed by NATIONAL BOOK NETWORK

Library of Congress Cataloging-in-Publication Data is available on file.

ISBN 978-0-7627-8475-2

Printed in United States of America

The author and Rowman & Littlefield assume no liability for accidents happening to, or injuries sustained by, readers who engage in the activities described in this book.

To my son Nelson, may he keep finding what he's looking for . . .

Contents

Acknowledgments

Special thanks to the museums that assisted me with this book.

Jim McReynolds, executive director of the Wallace Mining Museum in Wallace, Idaho, supplied several raw ore specimens from his excellent collection. His museum is a must-see when in Idaho's famed Silver Valley.

René Payne, image archivist at the Denver Museum of Nature and Science, patiently allowed me to browse her online image catalog and work through permission issues.

Dr. Lara O'Dwyer-Brown, curator of the Rice Northwest Museum of Rocks and Minerals in Hillsboro, Oregon, was always willing to rummage through her collection for another obscure sample.

Also, thanks to Rachel Houghton, veteran technical communicator and longtime friend from Portland, Oregon, who helped with photography, touch-up, editing, and encouragement, and to Martin Schippers, Frank Higgins, and Dirk Williams, who assisted in the field.

Rice Northwest Museum of Rocks and Minerals, ricenorthwestmuseum.org

Overview

About Geology

The term *geology* is a combination of two Greek expressions: "Geo" refers to the earth, and "logos" refers to the logic and language used to explain your observations. So think of geology as a way to organize and explain the earth processes that we see all around us. For some of the strange shapes we see, an educated explanation would be great. Most of what we know is good guesswork, based on lab experiments and inferences that take many detailed drawings to explain and a lifetime to understand. Fortunately, the more you see, the better things fall into place.

Geology is a young science, dating to 1815 if you start with William Smith's first geology map. But geology actually begins much earlier. Long ago ancient Greeks such as Pliny the Elder described various rocks and minerals, and many other scholars documented the metal mines of the times. Later, some of the smartest and most educated scientists laid down the basics, such as Nicolas Steno (1638–86), who observed that, most of the time, the rocks at the bottom of a cliff are older than the rocks at the top. He called that the Law of Superposition, and it helped explain how fossil seashells ended up on mountaintops. He also developed the Principle of Original Horizontality, which states that sedimentary rocks are usually deposited flat, although there can be local pinching, advancing, and other variations. Later, James Hutton presented *Theory of the Earth* in 1785, and Sir Charles Lyell wrote *Principles of Geology* in 1830. Arguments soon arose over the question of whether geology happened in slow, methodical processes or in short, catastrophic bursts. Once these distinguished thinkers realized the great age of Earth had nothing to do with Biblical teachings, they knew the answer was actually "both."

There are two key points to consider when trying to understand geology: time and entropy.

- **Time.** The earth is a very young planet and thus still very active. But it's also very old. Even though scientists have measured the earth at 4.6 billion years old, that's young in the context of a 20-billion-year-old universe. Given enough time, a lot can happen on a young, geologically active planet. We have earthquakes, volcanoes, and moving continents. The forces that boil up from the earth's magnetic core are a long way from burning out, and they are relentless. Some activities happen quickly, like tsunamis, and we have the video. Other forces take millions of years, leaving clues like all the mica flakes lined up in a schist. Good field observers can identify the obvious signs of things that seemed to happen before and apply those signs to the present and future.

- **Entropy.** Things fall apart all the time. Stuff happens. Storms rearrange coastlines and rework river channels. Earthquakes, volcanoes, windstorms, and floods all move mountains and leave scars that "heal." A rock balanced precariously atop another rock will not remain for long; eventually, it will shake loose. The earth is very efficient at recycling all that surface mayhem, hiding many clues. Mountains rise, then get ground down under glaciers and unrelenting rain. Tight chemical bonds that hold atoms together eventually weaken thanks to water, heat, pressure, and time. Oxygen in the air constantly rusts iron and dissolves minerals. Those forces are always at work and are easy to predict but hard to imagine sometimes. Try to picture the Yellowstone River under flooding conditions that happen once every 100,000 years. That's mayhem on a continental scale. Now imagine the resulting gravel bars as the river recedes from flood stage, and think about the possibility of being the first rockhound to check for Dryhead agate on those newly stirred gravels.

Given enough time, almost anything can happen, and it usually does. We rarely see these processes at the surface, and we can

Rocky Mountain National Park, north-central Colorado
PHOTO COURTESY OF BOB FANSHIER

only imagine what takes place at great depths. That's where the logic comes in. There is a lot of math, chemistry, physics, biology, and just general science involved in sorting out what's going on in the field. But you're mostly interested in what you can see and collect, so read on.

Think in Series

We don't get many absolutes in nature, so numbers, such as percentages of minerals present, help when thinking about crystal compositions. Just as there are probably no two snowflakes that are exactly alike, most granites differ in some way. Some basalts may have more iron and magnesium present, and some may have more feldspar. Some may have more magnesium and some more iron. You can't exactly tell without expensive equipment. And usually it doesn't really matter to that many decimals if you have a rough idea. You just want to collect the interesting forms, and you don't need a PhD in structural geology to dig out a seam of agate. You do need a hard hat, however.

In addition, rocks that start out as one thing can turn into several different rocks after billions of years. Consider the metamorphic progression below:

Increasing metamorphism ———→

Mudstone	Shale	Slate	Phyllite	Schist	Gneiss

Each stage along the way represents a different setting for heat and pressure that altered the preexisting rock. The boundaries may even be fuzzy or under dispute. Memorizing the series isn't the important thing—you just need to know it exists for right now.

Then there's the series just for schist:

Increasing metamorphism ———→

Chlorite	Biotite	Garnet	Staurolite	Kyanite	Sillimanite

Again, there are specific temperatures and other conditions scientists measured in lab experiments. There are reaction series, ranges of sediment sizes, and other different ways to explain rocks, minerals, and earth processes.

Field Studies

One big complication in identifying rocks and minerals is the tendency of the earth's atmosphere to oxidize everything. That same oxygen we need to breathe also wreaks havoc on fresh material. Oxygen ions are always looking to hook up with another ion, preferably a metal. You've seen how a freshly sharpened knife blade starts to rust in the rain. That oxidation is also at work on cliff faces and boulders, aiding and abetting a tendency toward natural cracks, root action, freezing, and thawing. Many rocks get a reddish-brown surface in a short amount of time. First they turn color and then they fall apart.

So the air might attack a fresh surface from the outside, while water flowing through cracks can alter the inside. Water is a great solvent—it tries to dissolve everything it meets. Then toss in the sandblasting effect of fierce winds carrying grit, plus roots breaking down cracks, spring floods, and extreme baking heat, and you can see that rocks at the surface will not last long. It can be hard to tell the difference between an old sandstone and a much younger basalt if they both weathered to the same rusty brown.

The Rocky Mountain States

For the purposes of this guide, the following states make up the Rocky Mountain states:

- Colorado
- Wyoming
- Idaho
- Montana

Each of these states offers unique collecting opportunities, and there is a rockhounding title in the FalconGuide series for each one. Use this book as a companion title to help identify what you find in the field or help you plan your next trip.

Artist's Point, Yellowstone National Park (No collecting!)

The topographic provinces for this region are easy enough to learn. The Rockies are the backbone of North America and stretch roughly north–south to dominate these four states. They contain ancient rocks dating to the Precambrian era and also host numerous granite batholiths. There are remnants of great inland seas to the east and west, with vast plains of sedimentary rocks, and signs of dramatic volcanic activity to the west, stretching across Idaho to the great Yellowstone caldera.

With dramatic mountains, enticing old mining camps, forbidding deserts, ancient rocks, and recent volcanism, there is a lot of geology to enjoy.

About This Guide

This guide was organized with beginners in mind. The most common rocks start first, with a table to help you narrow down what you're looking for, and pictures if you are still baffled. The thinking here was that if you don't know the name, organizing the contents alphabetically won't help much.

Because the Rocky Mountain states are rich in fossils, many quite valuable or scientifically interesting, there is a section for various fossils. When possible, you'll find information about fee-dig operations for certain specimens.

For the minerals, there are sections for common minerals and metals. The common minerals form most of our rocks and are rarely important to miners. The minerals in the metals section are generally valuable, come from mining districts, and are usually either attractive or economically interesting. Most of the highly scientific jargon and measurements stayed out; the information you need, like streak, hardness, and color, stayed in.

Finally, there are gemstones. These are the most valuable or highly sought rocks and minerals from the Rockies. You can try to get them on your own, using one of the FalconGuides books as a companion to this title, or you can haunt the shows and shops and online retailers. What you collect, and the order you collect in, is up to you.

Hardness

A lot of the time, especially when you're starting out, you can't always tell what you have. There are several ways to key out a rock or crystal, such as the streak, the hardness, the luster, the color, and the weight. I've provided a glossary at the end of this book to help you with many geological terms and expressions, and I've tried to keep the text as free as possible from the daunting vocabulary.

The hardness test is one of the most elementary ways to sort out samples. Try to keep a few of these common specimens on hand, and if possible, memorize the ten standard listings. I included the "nonstandard" testing materials in italics.

Mohs #	Material
1	Talc
2	Gypsum
$2\frac{1}{2}$	*Fingernail*
3	Calcite
$3\frac{1}{2}$	Copper penny
4	Fluorite
5	Apatite
$5\frac{1}{2}$	*Knife blade, nail, or window glass*
6	Orthoclase feldspar
$6\frac{1}{2}$	*Pyrite; steel file*
7	Quartz; agate, chert, chalcedony, and jasper; streak plate
$7\frac{1}{2}$	*Garnet*
8	Topaz
9	Corundum
10	Diamond

Be a Kind Collector

Before going any further, let's talk about safety and etiquette in the field. The biggest safety tip is to never push your car beyond its limits, and bring tools. Upgrade your tires, and be religious about vehicle maintenance. Your expeditions can take you to the end of some very long and dangerous roads. Walking out is no fun.

Don't treat private property like public lands; get all the permissions you need beforehand or give up and move on. Obey all signs, control your kids and dogs, pick up litter, and take only what you need when collecting. Consider bringing along an extra bucket to carry out some of the trash you may encounter when searching road cuts. Bring a shovel and use it. Bury toilet paper.

Additional Reference Materials

This book can help you with about 99 percent of the geology you'll see in the field, because most of the visible rocks at the surface of the earth are common sedimentary rocks. Igneous rocks are also extremely plentiful, and once you know the basic rules for those, you'll be able to screen them out as you look for nicer specimens.

National Audubon Society Field Guide to North American Rocks and Minerals is an 852-page treasure trove of 200+ color photographs. It's glossy, fits in a big pocket, and it probably takes a beating in the field. I leave mine at home.

For web-based research, I recommend Mindat, at mindat.org. It's a noncommercial online database of almost 39,000 mineral names, more than 300,000 photos, and 125,000 localities, and it is community editable. It has some GPS coordinates and can always use more.

The USGS operates the Mineral Resources Data System (MRDS) at http://tin.er.usgs.gov/mrds with excellent information on professional papers and publications related to mining districts, fossil locales, and more. It has interactive maps and plenty of data. The US Geological Survey map site at geocommunicator.gov has topo maps for the United States.

Another excellent online resource is Wikipedia, which has entries for many rocks and minerals. I also strongly recommend Andrew Alden's page at www.geology.about.com for a variety of general geologic information, including maps. Also, geology.com is a good site to skim around for info.

I'm a huge fan of the Roadside Geology series because they give even the most amateur geologist a general idea about the rock exposures passing by your windshield. I also recommend that every rockhound become familiar with geology maps, learn to use a GPS device, take great pictures, and share what you find via your favorite social networking site. See you out there!

Land Use

Some of the most lasting collectibles from the Rocky Mountain states come home via a camera rather than a rock bag. The vistas you'll encounter in national parks are normally composed of common rocks in dramatic formations, and they are worth every penny. However, it is important to understand the public land policies specific to national parks:

- No collecting of any kind is allowed. No pebbles as souvenirs, no yard rocks, no samples
- No camping except in designated areas
- No hiking except on official trails, sometimes requiring proper paperwork, such as backcountry permits
- No driving except on designated roads

Public lands are held in trust for all citizens to enjoy. The US Forest Service and the Bureau of Land Management are the primary land stewards in the Rocky Mountain states, while other agencies such as state, county, and wildlife agencies also control minor lands. Obey their rules and enjoy the scenery! For more information about the National Park Service, check their website at www.nps.gov.

State Designations

It helps to know what fossils, rocks, minerals, and gems are recognized as "official" for a state. Some regions have recognized more material, but here's a list so far for the Rocky Mountain states:

State	Fossil	Rock	Mineral	Gemstone
Colorado	Stegosaurus	Yule marble	Rhodochrosite	Aquamarine
Wyoming	Knightia fish			Nephrite jade
Idaho	Hagerman horse			Star garnet
Montana	Duck-billed dinosaur			Sapphire and agate

Celebrating the Wonders

The National Park System protects some of the most celebrated geological treasures in the Rocky Mountain States. The following venues belong on anyone's to-do list while traveling in that region:

Yellowstone National Park (Idaho, Montana, and Wyoming)
The most dangerous active caldera in North America shows geothermal forces at work.

Fossil Butte National Monument (Wyoming)
Eocene-age fossil fish and insects, with fee-dig opportunities on nearby private land.

Devil's Tower National Monument (Wyoming)
The first monument in the National Park System, designated in 1906. Columnar jointing on a massive scale.

Glacier National Park (Montana)
Stunning vistas of the northern Rockies.

Colorado National Monument (Colorado)
Sandstones, schists, and gneiss from the Colorado Plateau.

Great Sand Dunes National Park and Preserve (Colorado)
Stunning, 700-foot sand dunes are the tallest in North America.

Dinosaur National Monument (Utah and Colorado)
Dinosaur bone quarry hosts an impressive wall of bones in situ.

Florissant Fossil Beds National Monument (Colorado)
Awesome displays, and fee-dig opportunities for fossil insects.

Rocky Mountain National Park (Colorado)
Mountain terrain from the heart of the Rockies.

Hagerman Fossil Beds National Monument (Idaho)
Fossil mammals galore, including the famed Hagerman horse.

Craters of the Moon National Monument (Idaho)
Lava flows, cinders, vents, bombs, and more—an igneous petrologist's dream world.

I
ROCKS

Igneous Rocks: Extrusive

Use the matrix below as a starting point for understanding how igneous extrusive rocks are organized.

Matrix of Extrusive Igneous Rocks				
Name	Color	Grain Size	Composition	Useful Characteristics
Basalt	Dark	Fine or mixed	Low silica lava	Has no quartz
Andesite	Medium	Fine or mixed	Medium silica lava	Plagioclase with pyroxenes
Rhyolite	Light to medium	Fine or mixed	Very high silica	Quartz is common
Felsite*	Light	Fine or mixed	Medium to high silica	Rich in quartz and feldspar
Dacite	Light	Mixed	Medium silica	Plagioclase with hornblende
Pumice	Light	Fine	Sticky lava froth	Small bubbles
Obsidian	Dark	Fine	High silica	Glassy
Scoria	Dark	Fine	Runny lava froth	Large bubbles
Tuff	Light	Mixed	Andesite, rhyolite, or basalt	Ash fragments; sometimes welded

not covered in this guide

Basalt

Basalt flows at Craters of the Moon National Monument near Arco, Idaho

Group: Igneous; Extrusive
Mineralogy: Plagioclase feldspar, olivine, and pyroxene
Key test(s): Fine grained, often with microscopic crystals only; can form dramatic columns
Likely locale(s): Volcanic regions

Basalt occurs in many parts of the Rocky Mountain states. It appears gray or light black when fresh but quickly weathers to a tan, yellow, or brown surface. It can be difficult to identify because it tends to be fine grained, so look for clues in the deposit itself. Vesicles are common; these holes sometimes fill with agate, chalcedony, or opal, in small, round "blebs" that can weather out and are collectible. Basalt can also host zeolites, quartz veins, and calcite veins under the right conditions. Andesite and rhyolite are thick, viscous lavas that build up faster, while basalt is runny and can flow hundreds of miles. While still referred to as extrusive, basalt can form in large sills and cool slowly enough to create impressive polygons, referred to as columnar jointing. These colonnades are collectible as impressive towers.

Basalt is common in the Rocky Mountain states, especially across southern Idaho's Snake River plain, such as at Craters of the Moon National Monument.

Andesite

Common andesite is not as dark as basalt and has no flow lines.
PHOTO COURTESY OF THE RICE NORTHWEST MUSEUM OF ROCKS AND MINERALS

Group: Igneous; Extrusive
Mineralogy: Intermediate; feldspars, pyroxene, and/or hornblende
Key test(s): Fine grained with pyroxene and plagioclase
Likely locale(s): Volcanic mountain ranges

Andesite is a common form of lava, usually lighter in color than basalt, and it dominates many volcanic mountain ranges, as its higher silica content causes it to pile up better, as opposed to flood basalts that flow for miles. This rock is named after the Andes Mountains, where it is a key component of those volcanoes. Andesite often exhibits a platy structure, appearing in outcrop as though it were stacked, but that isn't a telltale clue. Andesite is the extrusive equivalent of diorite, matching that rock in chemical composition. This lava is typically fine grained, with small crystals of plagioclase feldspar (such as andesine), hornblende, pyroxene (such as augite or diopside), and biotite. It makes a fine decorative rock, weathering from its light-gray appearance when fresh to a dark-gray or black color.

Andesite occurs in volcanic terrain throughout the Rocky Mountain states.

Rhyolite

Yellowstone Canyon slices through a dramatic sequence of rhyolite flows.

Group: Igneous; Extrusive
Mineralogy: Quartz, feldspar, biotite, hornblende, with accessory magnetite
Key test(s): Quartz crystals in fine-grained matrix
Likely locale(s): Common in pyroclastic flows and caldera fill

Rhyolite is another extremely common igneous rock, so learning to distinguish it from andesite and basalt is worthwhile. Its high silica content usually means rhyolite is lighter in color. In general, rhyolite is light gray, but it can appear yellow, pale yellow, and from pale red to deeper red. Rhyolite can also form in welded, compacted bands with dramatic color varieties, sometimes called wonderstone, which is carvable and takes a polish. Rhyolite cools faster than basalt and usually appears to be composed of fine-grained but distinguishable crystals. It is sometimes glassy. Flow banding is common, and crystals often show alignment under a hand lens, as do any small vesicles present.

The most famed rhyolite locale is Yellowstone National Park, a resurgent caldera that represents one of the biggest volcanic dangers on Earth, capable of erupting 200 cubic miles of material.

Dacite

Dacite is a thick, viscous lava that forms domes and dikes.
PHOTO COURTESY OF DAVE TUCKER, WESTERN WASHINGTON UNIVERSITY, HTTP://NWGEOLOGY
.WORDPRESS.COM

Group: Igneous; Extrusive
Mineralogy: Plagioclase feldspar, quartz, biotite, hornblende, and augite
Key test(s): Hornblende and plagioclase in fine-grained matrix
Likely locale(s): Volcanic ranges; domes, dikes, or sills

In composition, dacite ranks between andesite and rhyolite, typically featuring much larger crystals and a coarse, patchy appearance. Dacite is not common, so it can be difficult to get familiar with it. The color is quite variable, ranging from white, gray, and rarely black to even pale red or brown, and rarely even to deeper reds and browns. The matrix frequently appears oxidized. Mineralogically, dacite contains the same components as granodiorite, with plagioclase feldspar, quartz, biotite, hornblende, and augite. Dacite usually occurs as ignimbrite or tuff, or as domes, dikes, and flows, and it can display some flow banding.

One example of dacite in the Rockies is the Lone Mountain dacite intrusion in Montana.

Pumice

Pumice is so frothy and full of air that it floats.
PHOTO COURTESY OF THE RICE NORTHWEST MUSEUM OF ROCKS AND MINERALS

Group: Igneous; Extrusive
Mineralogy: Andesitic to rhyolitic
Key test(s): Floats; soft
Likely locale(s): Volcanic mountain ranges

Look for pumice in known volcanic regions, especially any area with recent events. Pumice is light enough to float, thanks to its frothy, glassy texture, which traps air and makes each rock buoyant. Pumice has a unique texture and feel that makes it easy to identify, and it can show dramatic flow lines. Many violent pyroclastic events spew pumice with ash and other material, causing pumice fragments to show up in breccias, tuffs, and other ash deposits. Pumice rafts are also common after volcanic eruptions in a marine environment. Collectors don't need more than a sample or two for a complete collection, unless considering industrial uses for pumice, including yard rock, potting soil, and abrasives.

Pumice is common in younger volcanic regions such as Idaho and Wyoming.

Obsidian

Obsidian is most common in Wyoming and Idaho; the famed Yellowstone site is off-limits to collectors.

Group: Igneous; Extrusive
Mineralogy: Rhyolitic
Key test(s): Glassy luster, sharp edges, conchoidal fracture pattern
Likely locale(s): In rhyolite flows

Obsidian has no crystal structure—it chilled too fast for the atoms to align in a crystal lattice—so technically it is a glass, and thus a rock, not a mineral. It's an easy rock to identify because it has a glassy luster when freshly chipped. When weathered or rolled in streams and rivers, however, with no fresh surface, obsidian can resemble every other rhyolite pebble on the outside. Other than the classic shiny black luster, look for evidence of flow, such as banding. The familiar fracture pattern is "conchoidal" or shell-like; knappers have long used that tendency to break, predictably, to fashion amazing stone tools. The most common obsidian is jet black, but there are other colors and forms.

Throughout the western Basin and Range Province, rockhounds use the term "Apache tears" for fragments of obsidian formed in perlite and varnished by desert winds. The San Juan Mountains would be a good place to look in Colorado. Idaho's Snake River Plains contain numerous obsidian outcrops, including Big Southern Butte. Montana is limited, and Wyoming's most famous locale is inside Yellowstone National Park and uncollectable.

Scoria

Scoria is denser and more solid than common cinder but has a similar texture.
PHOTO COURTESY OF THE RICE NORTHWEST MUSEUM OF ROCKS AND MINERALS

Group: Igneous; Extrusive
Mineralogy: Usually similar to basalt
Key test(s): Rusty, like cinders. Numerous vesicles; heavier than pumice
Likely locale(s): Volcanic regions

Sometimes called cinder, scoria resembles pumice because both typically contain numerous air pockets, or vesicles. While pumice floats, however, scoria will sink in water. Where pumice is soft and easy to crush, scoria is usually harder and more rigid. The name comes from the Greek word for "rust," and the orange-red rusty color is a common characteristic of some forms of scoria. Thus, the name *scoria* refers to a rock with a certain color or texture, rather than a specific mineralogy. The tops of lava flows, where considerable froth sometimes takes place, can harden into scoria. Similarly, some frothy ejected material from eruptions can harden as scoria. Scoria is often used in landscaping and in barbecue grills.

Scoria is common in volcanic regions such as Idaho's Snake River plains. Cinder cones and scoria are both common around Big Southern Butte, for example.

Tuff

Most of the Painted Hills in the western deserts are tuff deposits.

Group: Igneous; Extrusive
Mineralogy: Basalt to rhyolite
Key test(s): Soft, unless welded; angular clasts
Likely locale(s): Downwind from volcanic ranges

Tuff, or lithified volcanic ash, is a common fragmental deposit around and downwind from volcanoes. Also called tephra, the term can be a bit of a catchall for the pyroclastic material ejected and deposited downwind. Tuff is usually tan to dark brown in color but can occur as pink, yellow, green, or even purple. There are varieties of tuff known as crystal tuff, containing mineral crystals; lithic tuff, with rock fragments; and welded tuff, in which pumice fragments were hot enough to compress in glassy wisps. Geologists also denote rhyolitic tuff, andesitic tuff, basaltic tuff, ultramafic tuff, trachyte tuff, and others. Tuff is usually found as a breccia, with a fine- to medium-grained matrix, and can contain large angular material such as pumice, volcanic bombs, and other debris. Tuff beds can be prime fossil-hunting locales, containing petrified wood, leaves, bones, and other organic debris.

Many of the "painted hills" in the western United States are tuffs and volcanic ash.

IGNEOUS ROCKS: INTRUSIVE

Below is a table comparing the various igneous intrusive rocks.

Matrix of Intrusive Igneous Rocks				
Name	Color	Grain Size	Composition	Features
Granite	Light, often pink	Coarse	Feldspar, quartz, mica, and hornblende	Wide range of color; coarse grain size
Syenite*	Light	Coarse	Mostly feldspar and minor mica	Like granite but no quartz
Tonalite*	Light to medium; salt-and-pepper	Coarse	Plagioclase feldspar and quartz plus dark minerals	Limited alkali feldspar
Porphyry	Any	Mixed	Large grains of feldspar, quartz, olivine, pyroxene	Large grains in a fine-grained matrix
Gabbro	Medium to dark	Coarse	High calcium	No quartz; limited olivine
Diorite	Medium to dark	Coarse	Low calcium	Limited quartz
Peridotite	Dark; often greenish	Coarse	Olivine present	Dense; 40+% olivine
Pyroxenite*	Dark	Coarse	Pyroxene	Rich in pyroxene
Dunite*	Green	Coarse	Olivine dominant	Dense; 90+% olivine
Kimberlite	Dark	Coarse	Source of diamonds	Found in "pipes"
Pegmatite	Any	Very coarse	Usually granitic	Dikes; small intrusions

Not covered in this guide

Granite

Granite can be pink, gray, or a mix called "salt-and-pepper."
PHOTO COURTESY OF THE RICE NORTHWEST MUSEUM OF ROCKS AND MINERALS

Group: Igneous; Intrusive
Mineralogy: Quartz, feldspar, hornblende, biotite, and magnetite
Key test(s): Often pinkish; exfoliation
Likely locale(s): Mountainous terrain

Granite is one of the most common and recognizable igneous intrusive rocks. It is usually coarse grained and is primarily composed of feldspar, hornblende, and quartz. It can be massive or display zoning, depending on how close it came to the exterior of the intrusion. It is sometimes a salt-and-pepper rock; some granites have a little more white feldspar, while others have a lot more dark hornblende. Granite may occur as minor intrusions, as larger plutons, or as huge batholiths taking up tens of thousands of square miles. One field clue is the tendency of granite to weather by exfoliation, where thin sheets peel off like the skin of an onion. Granite isn't highly collectible, but artisans cut granite into tabletops and counters or use granite in buildings and monuments because granite takes a high polish. Probably the most interesting aspects of granite are its association with economic ore deposits, often found along the margins of granite intrusions, and the presence of pegmatites.

The Idaho Batholith covers much of central Idaho; the Kaniksu Batholith occupies much of that state's Panhandle region. Colorado's Front Range is composed of the Pike's Peak Batholith, the Boulder Creek Batholith, and smaller plutons. The Boulder Batholith in Montana is associated with the rich mineralization near Butte and Helena. Wyoming contains the Sherman Batholith, the Boulder Mountains, the Wind River Range, and more.

Porphyry

Andesite porphyry showing large, dark crystals
PHOTO COURTESY OF THE RICE NORTHWEST MUSEUM OF ROCKS AND MINERALS

Group: Igneous; Intrusive
Mineralogy: Typically basalt or rhyolite
Key test(s): Contains large quartz or feldspar crystals
Likely locale(s): Volcanic regions

Rockhounds call a rock a *porphyry* when it appears to contain notable large crystals in a fine-grained matrix. The chemical composition of the large crystals and the matrix may be similar, but something occurred to get two different phases of cooling in these rocks. The large crystals are usually 2 millimeters or greater in size and are typically distinguishable as either quartz or feldspar. This apparently represents a time when the liquid magma cooled slowly, deep inside the earth, because large crystals typically take a long time to form. But subsequent events led to the still-liquid magma rising rapidly to the surface and then cooling quickly, so that the large crystals formed. Most igneous rocks seen in the field show this texture if studied closely, but striking porphyries will catch your eye.

This texture occurs in many volcanic rocks throughout the Rocky Mountain states.

Gabbro

Gabbro typically has a very coarse texture.
PHOTO COURTESY OF THE RICE NORTHWEST MUSEUM OF ROCKS AND MINERALS

Group: Igneous; Intrusive
Mineralogy: Feldspar, hornblende, biotite, and magnetite
Key test(s): Dark with coarse grains
Likely locale(s): Mountainous terrain

Gabbro is an interesting intrusive rock, distinguished by very coarse texture and a range of dark colors. One helpful characteristic is the presence of green olivine, which contrasts with the usual black or dark-gray color. Gabbros can be dark red, however, so that color test isn't precise. Gabbro sometimes makes up the base unit of massive, layered intrusions such as the famed Stillwater Complex in Montana. Perhaps the most famous layered intrusive with gabbro at the base is the storied Bushveld Complex of South Africa, the source of much of the world's platinum. These complex intrusions are often rich in chrome as well. Layering is sometimes visible in fresh exposures, but gabbro tends to weather faster and that can also aid in field identification. Other names include diabase, dolerite, and black granite. Gabbro takes a polish and can be fashioned into dark countertops, monuments, and statues.

The gabbros of Montana around the Stillwater Complex are the most noteworthy.

Diorite

Granodiorite from the northern Rockies
PHOTO COURTESY OF THE RICE NORTHWEST MUSEUM OF ROCKS AND MINERALS

Group: Igneous; Intrusive
Mineralogy: Feldspar, hornblende, biotite, and pyroxene
Key test(s): Dark smears, unlike granite; often in dikes and sills
Likely locale(s): Mountainous terrain

Diorite is a medium- to coarse-grained igneous intrusive rock, made up of common minerals such as plagioclase feldspar and pyroxene. It is usually medium grained, so crystals are recognizable under hand lens, but it can be coarse, with zoning common. Diorite is usually gray to dark gray, but it can be lighter, and a green or brown tint isn't out of the question. Salt-and-pepper colors are very common. Compared to granite and granodiorite, plain diorite is not common, and it often occurs as dikes, sills, and stocks at the margins of large granite batholiths. Because diorite is relatively hard, it takes a good polish. Ancient artisans used diorite for inscriptions, such as for the Code of Hammurabi, in a black diorite, and the Rosetta Stone, in a granodiorite.

Just about every major area that contains granite also contains diorite.

Peridotite

Chunky peridotite with visible olivine
PHOTO COURTESY OF THE RICE NORTHWEST MUSEUM OF ROCKS AND MINERALS

Group: Igneous; Intrusive
Mineralogy: Olivine and pyroxene
Key test(s): Green; coarse
Likely locale(s): Ancient terrains

Peridotite is a dark, dense rock made up of olivine and little else. *Peridot* is the gem variety of olivine and lends its name to this rock. Peridotite can be as much as 90 percent olivine; higher concentrations would make it a dunite. The olivine is usually present as large crystals, thus a coarse texture is another key distinguishing characteristic. It can form in layers or as massive structures. Peridotite is one of the most exotic rocks available in the field, as it is closely associated with the earth's interior and is probably the primary rock in the upper mantle. Since olivine reacts quickly with water and oxygen, peridotite is not very stable, and upon exposure to the atmosphere, it quickly starts converting to serpentinite.

Peridotite locales include Montana's Stillwater Complex and Wyoming's State Line kimberlite district.

Kimberlite

Sloan kimberlite from southeast Wyoming
PHOTO COURTESY OF W. DAN HAUSEL, SLOAN-KIMBERLITE.BLOGSPOT.COM

Group: Igneous; Intrusive
Mineralogy: Altered olivine, rare garnets
Key test(s): Very coarse; large crystals
Likely locale(s): Older continental crust

Kimberlite is one of only two rocks that contain commercial amounts of diamonds, and that makes it interesting for rock-hounds and prospectors as well as field geologists. The rock matrix is usually dark gray or dark blue but weathers yellow. It is very coarse, with large, rounded crystals easily visible. The presence of so many large, chunky crystals makes the rock look like a breccia. Kimberlite probably intrudes from deep in the mantle, and possibly accelerates to the surface of the earth at high speed. Because kimberlite extends at depth, it forms a narrow, vertical "pipe" and occurs primarily in cratons, a geological term for stable, continental rock. It tends to weather faster than its surrounding host does, and geologists learned to look for round depressions and small ponds or lakes in known kimberlite areas.

The primary locale in the Rockies to look for kimberlite pipes is the State Line district on the Wyoming-Colorado border. Hausel's website lists others.

Pegmatite

Pegmatite features large crystals of feldspar, mica, schorl, and other minerals.
PHOTO COURTESY OF THE RICE NORTHWEST MUSEUM OF ROCKS AND MINERALS

Group: Igneous; Intrusive
Mineralogy: Quartz, feldspars, altered olivine, rare garnets, orthopyroxene, and chrome diopside
Key test(s): Rare garnets; very coarse crystals
Likely locale(s): Older continental crust

Pegmatites sometimes host the most prized of gems, such as rubies, emeralds, and sapphires, so they are important to learn about because of their potential reward. The problem is that they get complicated quickly. The word refers to their coarse texture with large crystals, sometimes resembling a patchwork quilt. Pegmatites are usually found at the margins of large granite bodies and contain the same minerals, such as alkali feldspar and quartz. Other minerals include quartz, tourmaline, topaz, mica, apatite, lepidolite, and monazite. There are varieties, such as granite pegmatite and nepheline syenite pegmatite, but pegmatites are all usually white due to the amount of feldspar present. Pegmatites can also appear light yellow, tan, or even gray. Zoning is common, with vugs and cavities highly sought for large crystals of quartz, smoky quartz, topaz, and other prizes.

Famous pegmatite belts and outcrops in the Rockies include Idaho's Latah County; the Tobacco Root Mountains in Montana; Albany, Fremont, Goshen, and Natrona Counties in Wyoming; and Fremont, Gunnison, Larimer, and Park Counties in Colorado. Colorado's famed Mount Antero, known for large, gem-producing pegmatites, is in Chafee County.

METAMORPHIC ROCKS

Use the matrix below as a starting point for understanding how metamorphic rocks are organized.

Name	Hardness	Foliation	Grain Size	Color	Other
Soapstone	Very soft	Foliated	Fine	Light	Greasy
Slate	Soft	Foliated	Fine	Dark	Striking sound
Phyllite	Soft	Foliated	Fine	Dark	Shiny and crinkly
Serpentinite	Soft	Nonfoliated	Fine	Green	Shiny, mottled
Marble	Medium	Nonfoliated	Coarse	Light	Calcite or dolomite by acid test
Greenstone	Medium	Relics	Fine	Light	Common
Argillite	Hard	Relics	Fine	Mixed	Common
Mylonite*	Hard	Foliated	Coarse	Mixed	Crushed and deformed
Schist	Hard	Foliated	Coarse	Mixed	Large, deformed crystals
Gneiss	Hard	Foliated	Coarse	Mixed	Banded
Migmatite*	Hard	Foliated	Coarse	Mixed	Melted
Amphibolite	Hard	Foliated	Coarse	Dark	Hornblende
Hornfels	Hard	Nonfoliated	Fine or coarse	Dark	Dull, opaque
Eclogite*	Hard	Nonfoliated	Coarse	Red and green	Dense; garnet and pyroxene
Quartzite	Hard	Nonfoliated	Coarse	Light	Quartz—no fizzing

Not covered in this guide

Slate

Dark slate with small, brassy pyrite crystals
PHOTO COURTESY OF THE RICE NORTHWEST MUSEUM OF ROCKS AND MINERALS

Group: Metamorphic; regional
Mineralogy: Muscovite mica, chlorite; pyrite is a frequent accessory
Key test(s): Foliation; visible crystals; density
Likely locale(s): Metamorphic terrain

Increasing metamorphism ⟶

Mudstone	Shale	Slate	Phyllite	Schist	Gneiss

Slate started life as a sedimentary rock, most likely a mudstone, but intense heat and pressure caused the minerals to reorient themselves. At this level of metamorphism, the principal minerals are micas, such as muscovite and chlorite. Pyrite crystals and cubes, sometimes weathered to limonite, are common in some slates. Slate is sometimes confused with phyllite, but if you have visible mica flakes and the rock is crumbly and easy to break, it's a phyllite. Slate is usually dark gray, thanks to the presence of graphite from carbonaceous parent material. Slate makes great flagstones and building material and still shows up in high-end billiard tables.

Known areas for slate in the Rocky Mountain states include Colorado's Proterozoic Uncompahgre Formation and western Idaho's Triassic island arc metamorphic rocks.

Phyllite

Phyllite outcrop showing crumbly texture

Group: Metamorphic; regional
Mineralogy: Muscovite mica, chlorite; pyrite is a frequent accessory
Key test(s): Foliation; visible crystals; density
Likely locale(s): Metamorphic terrain

Increasing metamorphism ⟶

Mudstone	Shale	Slate	Phyllite	Schist	Gneiss

Phyllite underwent more heat and pressure than slate but has not yet hardened into a schist, so it tends to be crumbly and easy to crush by hand. It tends to be medium to dark gray but can turn black if enough carbon was present in the original sedimentary rock. It is mostly mica, like muscovite and chlorite, and the abundance of mica is what contributes to the foliation. Geologists term this kind of rock "friable" because it is so crumbly. Phyllite grades into schist, and the boundaries can be subtle.

Locales include Colorado's Proterozoic Uncompahgre Formation and western Idaho's Triassic island arc metamorphic rocks.

Schist

Schist with almandine garnets and muscovite mica from Emerald Creek, Idaho

Group: Metamorphic; regional
Mineralogy: Muscovite, biotite, kyanite, and glaucophane varieties
Key test(s): Foliated mica; garnets; density
Likely locale(s): Metamorphic terrain

Increasing metamorphism ———→

Mudstone	Shale	Slate	Phyllite	Schist	Gneiss

Schist is one of the more common metamorphic rocks, and it is relatively simple to identify. Outcrops often contain striking folds that look like wavy lines. Schist is usually medium to coarse grained, with mica being obvious. Gray and brown are common colors, but schist can run to white and yellow, depending on how much iron is present to stain the rock. The hardness of a typical schist is highly variable. It can range to 1 if considerable talc is present and to around 5 with more hornblende. If quartz is dominant, the rock's hardness can reach 7 or so. Chlorite and albite schists have undergone low-grade heat and pressure, while the presence of garnet and epidote represent medium-grade metamorphism. Further heat and pressure results in kyanite, a blue, bladed mineral, and then staurolite, noted for its twinning.

Schist is common throughout the metamorphic zones of the Rocky Mountain states.

Gneiss

Banded gneiss from Georgetown, Colorado
PHOTO COURTESY OF THE RICE NORTHWEST MUSEUM OF ROCKS AND MINERALS

Group: Metamorphic; regional
Mineralogy: Muscovite mica, chlorite; pyrite is a frequent accessory
Key test(s): Foliation; visible crystals; density
Likely locale(s): Metamorphic terrain

Increasing metamorphism ———→

Mudstone	Shale	Slate	Phyllite	Schist	Gneiss

Gneiss (pronounced NICE) is another extremely common metamorphic rock. In metamorphic terrain, it can make up most of the pebbles and cobbles of a river or creek. Gneiss is medium to coarse grained, and the minerals align tightly. This rock shows off light and dark banding in an infinite variety, but it is usually harder than schist and thus sometimes easily distinguished from those rocks. Note, however, that there is no acknowledged cutoff between schist and gneiss. Banding and layering are key characteristics; rocks and outcrops demonstrate folding on both large and small scales. Hardness is highly variable: Scratch tests show a range of 6 to 7 for hard, quartz/feldspar layers, down to 2½ to 3 for biotite-rich layers. Key components include orthoclase feldspar, quartz, biotite, and hornblende. Accessories include almandine garnet, corundum, and staurolite. Look for intense folding, light-and-dark banding, and hard, erosion-resistant rocks at the heart of mountain ranges.

Gneiss is common in old metamorphic terrains across the nonvolcanic mountain ranges of the Rockies—especially in Colorado.

Argillite

Massive argillite boulder from Missoula, Montana
PHOTO COURTESY OF THE RICE NORTHWEST MUSEUM OF ROCKS AND MINERALS

Group: Metamorphic; regional
Mineralogy: Muscovite mica, chlorite; pyrite is a frequent accessory
Key test(s): Foliation; visible crystals; density
Likely locale(s): Metamorphic terrain

Argillite is a common term for any fine-grained sedimentary rocks, such as siltstone, that have begun to metamorphose. Think of argillite as a mud or clay that has hardened into a rock. Argillites can be highly variable in composition but generally have high concentrations of aluminum and silica. Argillites sometimes mix with shale, making the distinction difficult. Argillite tends to be massive, but bands and zones of different composition give them an almost sedimentary look. Related rocks include black jadeite, catlinite, pipestone, dickite, and haidite.

One notable locale for argillite is the Superbelt area of northern Idaho and neighboring Montana, where it is common.

Amphibolite

Amphibolite is composed mostly of hornblende and is very dark.

Group: Metamorphic; regional
Mineralogy: Amphiboles such as hornblende and actinolite; feldspar
Key test(s): Crystalline often salt-and-pepper
Likely locale(s): Metamorphic terrain

Amphibolite is a general term for dark, dense metamorphic rocks that contain mostly members of the amphibole family. Geologists refer to an "amphibolite facies" to denote a particular pressure and temperature range required to produce the minerals inside. There may be bands or lines present, and it can be easy to confuse amphibolite with a banded gneiss. There is little or no quartz present in amphibolite. There are at least a dozen different amphiboles, but generally hornblende and actinolite are the two major minerals present. There are many different rock types that can metamorphose into amphibolite, including basalt and sedimentary rock. Note that garnets and rubies are often associated with amphibolite.

There are numerous amphibolite locales in the older rocks of all the Rocky Mountain states.

Greenstone

Greenstone is a sign of regional metamorphism.

Group: Metamorphic; regional
Mineralogy: Muscovite mica, chlorite; pyrite is a frequent accessory
Key test(s): Foliation; visible crystals; density
Likely locale(s): Metamorphic terrain

Greenstone is a dense, fine-grained rock that probably started life as a gabbro or, more likely, a seafloor lava flow such as pillow basalt or plain basalt. It has undergone regional metamorphism to cook into a plain, green appearance. It is generally massive, with occasional evidence of its basalt parentage, such as vesicles. It has a dull luster and is green, pale green, and occasionally yellow green. The hardness for greenstone is about 5 to 6, but it can vary depending on which minerals dominate the specimen. Prospectors throughout the Rockies quickly learned to locate greenstone outcrops as the telltale green color was easy to spot. Its presence signifies regional metamorphism, and finding it usually leads to discovery of quartz and calcite veins that could transport economic ore deposits. However, it can be very common in some areas, and it is not a sure indicator of economic ore deposits.

Greenstones were important in the gold-bearing South Pass, Rattlesnake Hills, and the Laramie Range areas of Wyoming; Montana's Tobacco Root Mountains; and the Pecos Greenstone Belt, which stretches into southern Colorado.

Hornfels

Spectacular lime hornfels with large biotite blades

Group: Metamorphic; regional
Mineralogy: Muscovite mica, chlorite; pyrite is a frequent accessory
Key test(s): Foliation; visible crystals; density
Likely locale(s): Metamorphic terrain

Hornfels is a common metamorphic rock in some terrains, but it can be difficult to identify. The name refers to the erosion-resistant Matterhorn in the Swiss Alps, where the rock forms prominent peaks. Biotite hornfels is usually black, or at best dark gray, while lime hornfels is lighter in color. Both varieties are very dense. Hornfels can be very fine grained and massive, with very few signs of relict bedding. Hardness ranges from 6 to 7, similar to gneiss. When pounded with a hammer, hornfels breaks into angular splinters, but that isn't a surefire field test.

The best places to look for hornfels are around the margins of big batholiths and intrusions, such as the Kaniksu Batholith and the Idaho Batholith, where there was plenty of heat.

Quartzite

Typical quartzite pebble, fracturing across bedding planes

Group: Metamorphic; regional
Mineralogy: Muscovite mica, chlorite; pyrite is a frequent accessory
Key test(s): Foliation; visible crystals; density
Likely locale(s): Metamorphic terrain

Quartzite is a common pebble in many streams and rivers because it is hard and strong. Quartzite has a vitreous luster and often occurs as a round, white rock, especially if pure. Other forms of quartzite can be light gray, and it can even run to pink or light brown. Quartzite is usually fine grained, but sometimes crystals are actually visible and the rock can be medium grained. Quartzite deposits are typically thick, massive units, resistant to erosion and forming prominent bluffs. Quartzite will break across bedding planes, not along them. The hardness is typical of quartz, at 7, but it can be softer if significant amounts of feldspar and calcite are present. This rock takes a nice polish and has some industrial applications including landscape rock, tiles, flooring, and even ballast. Some collectors confuse quartzite pebbles for agate because they share similarities, but reserve the term "agate" for translucent, banded quartz.

Quartzite pebbles and outcrops occur throughout the Rockies.

Marble

Banded marble from southeast Idaho (left). Brilliant white Yule Marble from Colorado (right).
BANDED MARBLE PHOTO COURTESY OF THE RICE NORTHWEST MUSEUM OF ROCKS AND MINERALS

Group: Metamorphic; regional
Mineralogy: Muscovite mica, chlorite; pyrite is a frequent accessory
Key test(s): Foliation; visible crystals; density
Likely locale(s): Metamorphic terrain

There are dozens of varieties of marble, but in general, it is simply a limestone or dolomite that has been highly metamorphosed, usually into a fine- to medium-grained specimen with a soft, vitreous luster. Marble is normally white, sometimes brilliantly so, but it is often darker thanks to black, green, red, pink, or even yellow staining from various impurities. It often contains veins of darker material that originated in the limestone bed as clay, sand, silt, or chert. This produces a lined or "marbled" look. Because limestone deposits are often quite large, it's logical that marble is usually massive, with foliation, banding, and streaking all common. Marble has a hardness of only about 3 but can range from 2 to 5. Calcite marble effervesces with cold hydrochloric acid. Chemists must first pulverize dolomitic marble and then heat the acid to get the expected fizzing action.

Marble occurs throughout Idaho in the south, central, and western regions of Idaho. The most famed deposit of marble in the Rockies is the Yule Marble, Colorado's state rock. It is found in the West Elk Mountains of Colorado and was used to create the stunning exterior of the Lincoln Memorial in Washington, D.C.

Serpentinite

Serpentinite is a jumble of several related green minerals.

Group: Metamorphic; regional
Mineralogy: Muscovite mica, chlorite; pyrite is a frequent accessory
Key test(s): Green, greasy, streaky, and soft
Likely locale(s): Metamorphic terrain

Serpentinite is a collective term for about twenty different magnesium iron phyllosilicates commonly found in varying percentages in most deposits. Serpentine is the most well-known constituent, but other common minerals include antigorite, lizardite, and chrysotile. Many of these minerals are amphibole asbestos, so use caution. This rock is usually green, with streaks of black, yellow, white, or gray. It is very dense, which distinguishes it from soapstone. The best field test is serpentinite's slick, greasy feel, which is a key characteristic. The structure is usually massive, if jumbled, and serpentinite is famous for "slickensides," where microfaulting and zones of movement result in polished, "slick" surfaces. The hardness for serpentinite is usually 4 or less on the Mohs' scale, sometimes as low as 2½, depending on how much hard quartz is present. Serpentinite usually occurs in small pods, lenses, and layers, with steep tilting common. It is carvable, and its varied hues make for interesting sculptures, even if it is harder than soapstone.

Western Idaho's Blue Mountains contain notable deposits of serpentinite.

SEDIMENTARY ROCKS

Use the matrix below as a starting point for understanding how sedimentary rocks are organized.

Name	Hardness	Grain Size	Composition	Features
Coal	Soft	Fine	Carbon	Black; burns
Shale	Soft	Fine	Clay minerals	Splits in layers
Limestone	Soft	Fine	Calcite	Fizzes
Dolomite	Soft	Coarse or fine	Dolomite	Fizzes if powdered
Phosphorite	Soft	Coarse or fine	Phosphorous	Gray or white
Coquina*	Soft	Coarse	Fossil shells	Bits and pieces
Wacke/ graywacke*	Hard or soft	Mixed	Mixed sediments with rock grains and clay	Gray or dark and dirty
Concretion	Hard, round	Fine	Lime-rich matrix; organic material inside	Round "cannonballs"
Conglomerate	Hard or soft	Mixed	Mixed rock and sediment	Rounded rocks glued together
Breccia	Hard or soft	Mixed	Mixed rocks and sediment	Sharp-edged conglomerate
Sandstone	Hard	Coarse	Clean quartz	Grainy
Arkose*	Hard	Coarse	Quartz and feldspar	Quartzy sandstone
Mudstone*	Soft	Fine	Clay-rich hardened mud	Hard to see particles
Siltstone	Hard	Fine	Very fine sand; no clay	Gritty
Chert	Hard	Fine	Chalcedony	No acid fizz

*Not covered in detail in this guide

Coal

Common coal has a dark, dull to shiny appearance; the lower the grade, the messier it is.

Group: Sedimentary
Mineralogy: Carbon
Key test(s): Oily smell; dusty; light heft
Likely locale(s): Metamorphic terrain

Increasing metamorphism ———→

Peat	Lignite	Bituminous	Anthracite	Graphite	Carbon

Coal actually spans the boundary between a sedimentary rock and a metamorphic rock. Even at high grades, coal is very soft; its overall hardness ranges from 1 to 2½. Once coal bakes into the anthracite stage, it is less greasy, but coal dust is always a problem due to its tendency to be brittle and friable. Further metamorphism leads to graphite and pure carbon, but diamonds exist in kimberlite, not from metamorphosed coal. The higher the quality, the fewer relicts are present, and color can get more interesting, with purple iridescence. High-grade coal is up to 98 percent hydrocarbons, but sulfur and nitrogen are usually present.

Coal is rare in Idaho; there are scattered, limited deposits in Montana. Colorado has major coal beds along the western slope. Wyoming hosts numerous world-class coal deposits, especially in Campbell County, in the northeastern part of the state.

Shale

Shale from the Gros Ventre Formation in Wyoming

Group: Sedimentary
Mineralogy: Muscovite mica, chlorite; pyrite is a frequent accessory
Key test(s): Foliation; visible crystals; density
Likely locale(s): Sedimentary basins

Increasing metamorphism ⟶

Mudstone	Shale	Slate	Phyllite	Schist	Gneiss

Shale is a sedimentary rock, but it has started to harden in distinguishable ways. Varieties include oil shale, calcareous shale, carbonaceous shale, and others. Shale is usually dark but not always, as its color depends on the source rock from which it hardened. Shale is noted for the way it fractures into plates, along bedding planes, where evidence of past water action such as waves, ripples, cracks, and footprints are all observable. Shale beds are easy to spot, although they can easily be confused with platy andesite. Look for evidence of fossils or water features for one differentiator. Note that some rockhounds use the slang "shale" for unwanted rock, no matter what its true composition.

Shale is common throughout the Rocky Mountain states. For example, the Pierre Shale, which dates to the Cretaceous era, stretches from Canada to New Mexico and is famous for fossil concretions, oil deposits, and natural gas. It is typically dark gray and records part of the great Western Interior Seaway.

Limestone

Fossiliferous limestone from Wyoming (left) and banded travertine from near Beulah in Pueblo County, Colorado (right)
TRAVERTINE PHOTO COURTESY OF THE IMAGE ARCHIVES, DENVER MUSEUM OF NATURE AND SCIENCE

Group: Sedimentary
Mineralogy: Chemical deposition
Key test(s): Fizz test; fossils
Likely locale(s): Old sea basins

Limestone is one of the most common sedimentary rocks, and it comes in many varieties. Chalk is fine grained, derived from the skeletons of tiny sea creatures, while coquina contains large, abundant shell fragments. Travertine is typically banded and colorful, while oolitic limestone refers to tiny orbs, or oolites, which are very small concretions. The term "marl" describes a limestone with a high percentage of silicates. Limestone is often marked in the field by pitted and pockmarked outcrops. The hardness for limestone is in the range of 3 to 4. The most reliable test is fizzing in cold, diluted hydrochloric acid. Typically, limestones are light gray, but if iron is present, they can stain reddish, trending to darker gray.

Limestone is common throughout the Rocky Mountain states. Lewis and Clark Cavern in Montana sits in limestone; so does Minnetonka Cave in southeast Idaho. Cave of the Winds near Pikes Peak, Colorado, is another famous limestone cave.

Dolomite

Common dolomite from the Bighorn Formation of Wyoming

Group: Sedimentary
Mineralogy: Chemical
Key test(s): Fizz test
Likely locale(s): Common

Dolomite, sometimes called dolostone to differentiate it from crystalline dolomite, is typically tan or light gray, but varieties can range all the way to pink and dark gray. It is typically dense like limestone, its cousin, but there is usually less evidence of grains or fossils. Typically, dolomite has a very fine texture because it doesn't have the shell fragments or oolites of limestone. Limestone and dolomite share many characteristics, but the big difference is that dolomite has substituted more magnesium for calcium. It can appear as bands and grade slowly into limestone. The best evidence for separating dolomite from limestone is with hydrochloric acid: Dolomite will only fizz if the acid is hot and the material has been ground into a powder, while limestone is much more reactive. Dolomite ends up as fertilizers high in calcium and magnesium.

Dolomite is rare in Montana, but it is common in southeast Idaho, Wyoming, and Colorado, especially adjacent to limestone rocks.

Phosphorite

Phosphate-rich phosphorite from the Phosphoria Formation of southeast Idaho and western Wyoming

Group: Sedimentary
Mineralogy: Chemical
Key test(s): Fizz test
Likely locale(s): Sedimentary basins

Phosphorite is typically tan or dark white. It is typically dense like dolomite (also called dolostone), but there is usually less evidence of grains or fossils. Phosphorite has a very fine texture because it doesn't have the shell fragments or oolites of limestone. Phosphorite and dolomite share many characteristics, but the big difference is that dolomite has substituted more magnesium for calcium. It can appear as bands and grade slowly into shale or dolomite. The best evidence for separating dolomite from phosphorite is with hydrochloric acid: dolomite will fizz if the acid is hot and the material has been ground into a powder.

Phosphorite from the Phosphoria Formation in southeast Idaho and western Wyoming supplies most of the phosphate fertilizer in the United States.

Concretions

Large Frontier Formation ammonite just emerging from concretion (left) and prepped, out of the Bearpaw Formation, in Montana (center); iron oxide nodules from Hahn's Peak in Routt County, Colorado (right)
AMMONITE CONCRETION PHOTO COURTESY OF BRUCE THIEL, NORTH AMERICA RESEARCH GROUP, WWW.NARG-ONLINE.COM. PREPPED AMMONITE PHOTO COURTESY OF RON FORTIER, WWW.PREHISTORICALBERTA.LEFORA.COM. IRON CONCRETIONS PHOTO COURTESY OF THE IMAGE ARCHIVES, DENVER MUSEUM OF NATURE AND SCIENCE

Group: Sedimentary
Mineralogy: Clastic environment usually
Key test(s): Round, heavy
Likely locale(s): Sandstone

Like geodes, concretions are like a lottery ticket of the geology world. The outside appearance may be dull and drab, but inside may contain marvelous treasures. Concretions are typically round or oval structures that can range from tiny, pea-size orbs to large, beach ball–size or larger monsters. A concretion typically forms when some organic debris starts rolling around in a lime-rich mud within an active marine bay or lagoon. The mud will stick to organic material and soon forms a small, round ball. Inside could be a fossil or, in some cases, a septarian nodule with minerals such as barite. Or iron oxide could replace the calcium carbonate, creating an iron oxide nodule.

Experienced fossil hunters know the Pierre Shale contains ammonites within its concretions—some of which are beach ball–size or larger. The Pierre Shale (and equivalents, such as the Frontier Formation and Bearpaw Formation) stretches across much of Montana, Wyoming, and Colorado. The Glendive, Montana, area hosts ammonites and also septarian nodules, which are complex concretions that have formed, dried, cracked, and refilled, sometimes with barite. Around Lovell, Wyoming, rockhounds collect dahlite nodules, an obscure carbonate-apatite sphere that contains radiating crystals inside.

Conglomerate

Conglomerate contains rounded pebbles, often with a very strong matrix.
PHOTO COURTESY OF THE RICE NORTHWEST MUSEUM OF ROCKS AND MINERALS

Group: Sedimentary
Mineralogy: Large clasts in cement matrix
Key test(s): Rounded pebbles
Likely locale(s): Nonmarine

Conglomerates typically occur as hard, haphazardly sorted beds of large, rounded pebbles glued together in a fine-to medium-grained matrix. If the pebbles are still angular and edgy, the geologists call the rock a breccia. This is a clastic rock, and the clasts themselves can be big or small. By definition, the rocks and pebbles are greater than 0.08 inch in diameter, but they can range to large cobbles and boulders. Because conglomerates typically form in nonmarine waters that are very active, the clasts are usually hard material such as quartzite or chert, which can survive that much churning. The matrix of a conglomerate is often silica or calcite, sometimes making this rock extremely tough and resistant to erosion. Fossils are rare in conglomerates because the high velocity required to move these rocks could destroy anything soft or fragile.

Conglomerates are very common across the Rocky Mountain states.

Breccia

Breccia is composed of angular clasts that have not had their corners rounded off.

Group: Sedimentary
Mineralogy: Large clasts in cement matrix
Key test(s): Rounded pebbles
Likely locale(s): Nonmarine

Unlike conglomerates, breccias consist of broken rock pieces that are still angular and sharp edged. This is usually evidence that the source material is located nearby, as significant rounding hasn't started. These rocks are similar to tuff and volcanic ash and are usually lighter in color. They contain mixed material, but the general rule is that the rock fragments are coarse, angular, and varied. Breccias often occur in thick, massive beds, possibly interbedded with fine-grained tuff. Breccias are sometimes evidence of explosive forces, such as meteorite-impact breccias, and from volcanic explosions. Breccias can also depict evidence of significant movement along faults.

Volcanic and sedimentary breccias are common across the Rocky Mountain states.

Sandstone

Red sandstone with ripple marks, from the Flathead Formation of Wyoming

Group: Sedimentary; clastic
Mineralogy: Large quartz or feldspar clasts in cement matrix
Key test(s): Rounded pebbles
Likely locale(s): Nonmarine

Sandstone is one of the most common sedimentary rocks, and it occurs throughout the region. Frequently, sandstones in the field are interbedded with mudstone, limestone, shale, and other sedimentary rocks. By definition, all sandstones are marked by a grain size of 0.05 to 2.0 millimeters, which makes it medium grained compared to mudstone and siltstone but not coarse like a conglomerate. Sandstone displays great variety in color, occurring in hues of gray, brown, red, yellow, and even white. What holds the clasts or grains together in a sandstone is usually silica, but calcite or even iron oxide will also serve. Sandstone is relatively easy to shape and fashion into buildings and walls, and it is often used as a decorative stone.

Sandstones from the Cretaceous era are common in the Rocky Mountain states, and locally they contain abundant dinosaur fossils. Two of the more famous formations are the Morrison Formation and the Hell Creek Formation.

Siltstone

Siltstone with fossil trilobite, from Delta, Utah
PHOTO COURTESY OF THE RICE NORTHWEST MUSEUM OF ROCKS AND MINERALS

Group: Sedimentary; clastic
Mineralogy: Very fine
Key test(s): Visible bedding planes
Likely locale(s): Marine

Siltstone is a common sedimentary rock that is light gray or brown in color, but it can be darker if more organic material or iron staining is present. It is usually dense, and under a microscope it contains small particles of silt or clay. Bedding planes are often easy to see, especially if wet. Look for clay, such as kaolinite, feldspars, and quartz grains, plus mica flakes, but at a very small scale. Fossils tend to be soft in these rocks and often require an immediate coating of polyvinyl acetate.

Siltstone is common across the Rockies, grading into sandstone, mudstone, and coarser rock units.

Chert

Chert inclusion found in limestone from Wyoming

Group: Sedimentary
Mineralogy: Silica
Key test(s): Hardness 7. Organic nature; ooze origins
Likely locale(s): Nodules in limestone

Like most quartz incarnations, chert has a hardness of 7. It is commonly white in color, but banded chert can take on reds, yellows, and even greens. Chert is a by-product of organic ooze from the seafloor hardening into a rock, while jasper results from circulating silica solutions, usually in basalt. At high magnification, some cherts display tiny little skeletons. Flint is a form of chert occurring as black or dull-gray chunks and nodules. Chert is dense, and smooth when polished, but can feel very rough where exposed. Bedded cherts can be big enough to slice with a rock saw and polish into cabochons.

Chert is common in the Phosphoria Formation of southern Idaho and western Wyoming, and among south and central Idaho's older limestone formations.

II
FOSSILS

Common Fossils

Fossil Algae (stromatolites)

Sliced and polished stromatolites from the Green River Formation near Wamsutter, Wyoming
PHOTO COURTESY OF THE RICE NORTHWEST MUSEUM OF ROCKS AND MINERALS

Group: Invertebrates
Mineralogy: Usually replaced by quartz
Key test(s): Visible wavy lines
Likely locale(s): Marine

Stromatolites are fossilized algae mats, representing the evolutionary advance from single one-celled organisms to colonies. They are among the oldest forms of life on earth, dating to the Precambrian era, and they still exist today. They can occur as nodules and round blobs or as wavy lines, and colors range from tan to gray and even purple. Some forms erode quickly, but hard pieces will take a polish.

Stromatolites are common across the Belt Supergroup from Idaho's Panhandle region to Libby, Montana, particularly near Glacier National Park. Wyoming also boasts collectible specimens, from the South Pass area, and younger Eocene specimens in the Green River Formation.

Fossil shells

Silicified freshwater snails from Wamsutter, Wyoming
PHOTO COURTESY OF THE RICE NORTHWEST MUSEUM OF ROCKS AND MINERALS

Group: Invertebrates
Mineralogy: Shells replaced by quartz
Key test(s): Hardness
Likely locale(s): Marine

Fossil shells occur in sedimentary rocks throughout the Rocky Mountain states, owing to the deposits left behind from great inland seas and huge freshwater lakes. Brachiopods are common in the limestones of Idaho and western Wyoming. Cretaceous clams occupy parts of southeast Colorado. Some of the most prized fossil shell specimens are from Delaney Rim, south of Wamsutter, Wyoming, where silica replaced the calcium carbonate in the shells, forming a type of chalcedony. The name that stuck was "turritella" but these are freshwater snails, whereas true turritella are marine. Paleontologists first changed the classification to *goniobasis* but now assign these snails as *Elimia*. The beds are remnants of ancient Lake Gosiute and are part of the Green River Formation.

Fossil leaves

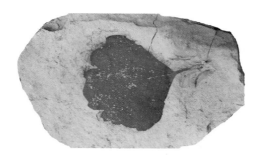

Beautifully preserved fossil leaf Trochodendraceae zizyphoides *from Clarkia, Idaho*
PHOTO COURTESY OF MARGARET L. JOHNSON, NORTH AMERICA RESEARCH GROUP,
NARG-ONLINE.COM

Group: Invertebrate
Mineralogy: Carbon remains
Key test(s): Appearance
Likely locale(s): Freshwater or marine rocks

Unlike collection vertebrate fossils, there are very few restrictions on collecting fossil leaves, and collectors can specialize in particular types, periods, ecosystems, and more. In addition, some fossil leaf locales, or quarries, are incredibly abundant, with great variety. Many of the siltstone, sandstone, and mudstone formations of the Rocky Mountain states feature fossil leaf material.

The Fossil Bowl fee-dig site at Clarkia, Idaho, boasts excellent preservation quality in its specimens. Called a *Lagerstätte*, German for "storage place," the site yields specimens that still contain DNA, and boast their original color when first exposed to oxygen.

Fossil wood (petrified wood)

Petrified wood from the Blue Forest, near Farson, Wyoming
PHOTO COURTESY OF THE RICE NORTHWEST MUSEUM OF ROCKS AND MINERALS

Group: Invertebrate
Mineralogy: Usually replaced by quartz
Key test(s): Visible wood structure, esp. growth rings
Likely locale(s): Primarily sedimentary; also ash deposits

Petrified wood is the common term for mineralized wood material. The mineralization is some form of silica, or quartz, including agate, chalcedony, opal, or jasper. Some specimens could have two or more forms of quartz, so the term "petrified" is still useful.

Petrified wood is quite common across the Rocky Mountain states, especially in Tertiary ash beds across southern Idaho, eastern Montana, and several places in Wyoming and Colorado. The Blue Forest near Farson, Wyoming, is a famed locale frequented by rockhounds.

Fossil bugs and birds

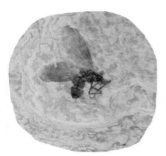

Fossil wasp from Fossil Butte National Monument, near Kemmerer, Wyoming (left). Fossil bird from the same locale (right).
WASP PHOTO COURTESY OF THE RICE NORTHWEST MUSEUM OF ROCKS AND MINERALS

Group: Invertebrate and vertebrate
Mineralogy: Insect hard parts usually replaced by carbon
Key test(s): Visible insect parts, such as wings and legs
Likely locale(s): Delicate thin-bedded sedimentary rocks

Fossil insects are rare, requiring delicate conditions and gentle ash, silt, or mud deposits. Worldwide, most well-preserved insects come from amber, but there are only minute amber deposits from Montana and Wyoming in the Rockies, associated with coal deposits. So insect hunters in the Rocky Mountain states must search sedimentary rocks, particularly from ancient lakes. Unlike petrified wood, where quartz is required, fossil insects are often carbonized or turned to charcoal.

Fossil Bowl is a reasonable fee-dig site at Clarkia, Idaho, and contains insect fossils from the Miocene era. Fossil Butte National Monument near Kemmerer, Wyoming, is famous for the variety of its Eocene-age insect population. Florissant Fossil Beds National Monument in Colorado contains abundant Eocene insect fossils—more than 1,500 separate species to date. Florissant sponsors digs during the summer.

Fossil fish

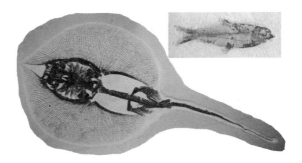

Fossil stingrays (left) are a challenge to prepare. The smaller fossil fish Knightia oceana *from the Green River Formation is Wyoming's state fossil.*
FOSSIL STINGRAY PHOTO COURTESY OF THE RICE NW MUSEUM OF ROCKS AND MINERALS.
KNIGHTIA PHOTO COURTESY OF TIM FISHER, OREROCKON.COM

Group: Vertebrate
Mineralogy: Usually replaced by quartz
Key test(s): Bone structure
Likely locale(s): Marine and freshwater sedimentary rocks

The fossil bones of fish are protected by law, as are all vertebrates, and only licensed paleontologists connected to museums can collect such fossils from public land. Private land is a different story, and there are many fee-dig operations across the Rocky Mountain states for that reason. The fossil fish *Knightia* was the equivalent of herring and sardines of today and was an abundant food source in much of the Green River Formation. It is Wyoming's state fossil.

Excellent fee-based fossil quarries surround Fossil Butte National Monument near Kemmerer, Wyoming. These quarries yield Eocene-age fish from thinly bedded limestone. For example, check out Ulrich's Fossil Gallery at ulrichsfossilgallery.com. There are others to search out as well.

Fossil mammals

Fossil mammoth skull from Hagerman Fossil Beds in Idaho

Group: Vertebrate
Mineralogy: Usually replaced by quartz
Key test(s): Bone structure
Likely locale(s): Marine and freshwater sedimentary rocks

When alive, mammal bone contains the mineral apatite and colla-gen tissue; upon death the collagen decays. When fossilized, silica replaces the collagen or occupies the voids left behind, making the bone hard and strong. As with all vertebrate fossil collecting, you should mark a spot you suspect contains mammal fossils and show it to a nearby museum curator or university paleontologist.

The sedimentary rocks throughout the Rocky Mountain states contain numerous mammal fossils. Since mammals didn't really take off until dinosaurs died away, look for fossil mammal bones in Cenozoic sediments. The famed fossil mammal locale at the Hagerman Fossil Beds in Idaho contains zebras, horses, mam-moths, and more. More recently a huge fossil mammal locale at Ziegler Reservoir in Colorado yielded Columbian mammoth, American mastodon, Ice Age bison, giant ground sloth, and a deerlike animal.

Fossil reptiles

Tyrannosaurus rex *skull; sliced and polished dinosaur bone*

Group: Vertebrates
Mineralogy: Usually replaced by quartz
Key test(s): Bone structure
Likely locale(s): Marine and freshwater sedimentary rocks

The first dinosaur fossil excavated in the United States came from north-central Montana in 1855. In the years that followed, a kind of "bone wars" broke out between leading paleontologists, as they competed to find and name as many species as possible. The Mesozoic sedimentary rocks of the Rocky Mountain states yielded thousands of specimens, and fee-dig opportunities still exist for aspiring dino-diggers. Hell Creek, Montana, is just one example (paleotrek.com).

Numerous excellent museums containing dinosaur fossils dot the Rockies. The best book on the subject for beginners, by far, is *Cruisin' the Fossil Freeway*, wonderfully written by Dr. Kirk Johnson, former chief curator at the Denver Museum of Nature and Science, and beautifully illustrated by Ray Troll, a noted, whimsical artist. Their wall map of the western United States is a must-have for aspiring paleontologists.

III
MINERALS

Common Minerals

Actinolite

A large specimen of actinolite and uralite, a mass of crystals with epidote from the Calumet Mine in the Turret district near Salida in Chaffee County, Colorado
PHOTO COURTESY OF THE IMAGE ARCHIVES, DENVER MUSEUM OF NATURE AND SCIENCE

$Ca_2(Mg,Fe)_5Si_8O_{22}(OH)_2$
Family: Amphibole silicates
Mohs: 5–6
Specific gravity: ~3
Key test(s): Green, fibrous, brittle
Likely locale(s): Metamorphic terrain

Actinolite is usually green, ranging from pale green to dark green, sometimes with yellow tinting. It can display as long, matted blades, usually quite brittle, and the blades can be parallel or create spectacular sprays. Like so many other minerals, actinolite occurs in a series, in this case with one end being rich in magnesium and the other rich in iron, and near-infinite percentages between the two extremes. Actinolite sits in the middle; magnesium-rich actinolite is tremolite, while iron-rich actinolite gets the unimaginative name *ferro-actinolite*. These rocks are a source of asbestos, so beware. Related rocks include nephrite jade.

   Actinolite is common across the Rocky Mountain states.

Agate–Cream

Cream agate from Mackay, Idaho
PHOTO COURTESY OF THE RICE NORTHWEST MUSEUM OF ROCKS AND MINERALS

Quartz, SiO$_2$
Family: Cryptocrystalline quartz
Mohs: $6\frac{1}{2}$–7
Specific gravity: 2.65
Key test(s): Banding; translucence
Likely locale(s): Veins and blebs in basalt

Agate is a term loosely used by rockhounds to describe any clear or translucent quartz, but the term is interchangeable with chalcedony, and chalcedony is actually the preferred term among mineralogists. However, agate hunting, agate picking, and agate digging are such a part of our lexicon that the term will never go away. Agate is one of the most popular and collectible forms of quartz, and it comes in a wide variety, including lace agate, moss agate, banded agate, and others. Usually the term "agate" denotes translucent, banded quartz, and usually the banding is quite distinctive. Typical colors are clear, red, yellow, or a pleasing, highly sought light blue. True agate has a fine grain and excellent color.

Mackay, Idaho, contains notable cream agate with dendrites and branching inclusions and little banding.

Agate–Dendritic

Translucent agate with dendritic inclusions from Crane, Montana, on the Yellowstone River in Richland County, Montana
PHOTO COURTESY OF THE RICE NORTHWEST MUSEUM OF ROCKS AND MINERALS

Quartz, SiO$_2$
Family: Cryptocrystalline quartz
Mohs: $6\frac{1}{2}$–7
Specific gravity: 2.65
Key test(s): Banding; translucence
Likely locale(s): Veins and blebs in basalt

Rockhounds use the terms dendritic agate or "moss" agate, especially when the dark dendrites are thick and banded. Dendrites are usually black, formed from manganese oxide, and often display interesting patterns and inclusions. Often rockhounds won't know if there are dendritic patterns until slicing and polishing their material. The "rind" or outside of such agates tends to be plain as well, so finding this kind of agate is difficult.

Southwestern Montana contains notable deposits of dendritic agate. Wyoming's Platte County also contains excellent dendritic agate. Other famed agate locales in the Rocky Mountain states include fluorescent Sweetwater agates in Fremont County, Wyoming. The famed agate beds at Graveyard Point, reached via Marsing, Idaho, yield moss gate. Grand and Larimer Counties in Colorado yielded moss agate in the past. Colorado's Weld County contains agate similar to the famed beds at Fairburn, South Dakota.

Agate—"Dryhead"

Dryhead agate from Montana's Yellowstone River
PHOTO COURTESY OF THE RICE NORTHWEST MUSEUM OF ROCKS AND MINERALS

Quartz, SiO$_2$
Family: Cryptocrystalline quartz
Mohs: 6½–7
Specific gravity: 2.65
Key test(s): Banding; translucence
Likely locale(s): Veins and blebs in basalt

Unlike common agate and chalcedony, Dryhead agate contains interesting banding patterns that may reflect repeated saturation by silica solutions as the material formed. Inconsistent chemical impurities such as iron give the bands distinctive colors and result in outstanding patterns when cut and polished.

Montana's Yellowstone River still contains notable pieces of Dryhead agate.

Amethyst

Loose purple amethyst crystals from Crystal Park in Beaverhead County, Montana, showing sharp termination
PHOTO COURTESY OF THE RICE NORTHWEST MUSEUM OF ROCKS AND MINERALS

Quartz, SiO$_2$
Family: Crystalline quartz
Mohs: 7
Specific gravity: 2.65
Key test(s): Purple; harder than fluorite
Likely locale(s): Quartz veins

Amethyst is a form of crystalline quartz, like smoky quartz, only amethyst is typically purple, sometimes vividly violet. Iron impurities impart the purple color. Like common quartz, amethyst reaches a hardness of 7 on the Mohs' scale. Crystals are hexagonal and often sharply terminated.

The Pohndorf Amethyst Mine in Jefferson County, Montana, produced notable amethyst plates and crystals with excellent terminations; miners in Silver Bow County also yielded good amethyst. Colorado's mines probably offer the most opportunities for amethyst, including the Empire district, which produced crystals of striking deep color, and the Ouray district. The mines near Creede also contained good amethyst. Idaho, including the Big Lost River area, offers opportunities for crystal-lined geodes that contain amethyst.

Apatite

Well-formed apatite crystal
PHOTO COURTESY OF THE RICE NORTHWEST MUSEUM OF ROCKS AND MINERALS

Calcium phosphate, Ca$_5$(PO$_4$)$_3$(F,Cl,OH)
Family: Phosphates
Mohs: 5
Specific gravity: 3.1–3.2
Key test(s): Hardness test
Likely locale(s): Pegmatites

Apatite is the name given to a group of phosphate minerals such as fluorapatite and chlorapatite, so the chemical formula can vary according to how much of one element or another is present. As expected, the color varies. Darker green, purple, or violet are common, as are red, yellow, and pink. Apatite is a common mineral in igneous rocks; larger crystals occur in pegmatites. Fluorite has similar coloration but isn't as hard, and the crystal habit is cubic, while quartz is harder. Interestingly, apatite is the hardening agent for bones and teeth.

Idaho's Orofino district and the Sawtooths yielded apatite in the past. Montana contains numerous apatite locales, including Yogo Gulch and the Butte area. In Wyoming, Fremont County and Big Horn County host apatite, as well as numerous locales north of Rock Springs. Colorado lists almost thirty good apatite locales in the literature, scattered across the state.

Aragonite

Aragonite needles are another form of calcium carbonate.
PHOTO COURTESY OF THE RICE NORTHWEST MUSEUM OF ROCKS AND MINERALS

Calcium carbonate, CaCO$_3$
Family: Carbonates
Mohs: $3\frac{1}{2}$–4
Key test(s): Fibrous appearance
Likely locale(s): Fossil shells, stalactites

Aragonite is another form of calcium carbonate, but unlike its related cousin calcite, aragonite usually occurs thanks to biological processes. Crystals are usually white or milky white, but with impurities, a vast range of colors are possible. Some mollusks form their shells completely with aragonite; other mollusks alternate between calcite and aragonite. Aragonite is not strictly biological, however. For example, the stalactites of Carlsbad Caverns are aragonite. The flowering pattern is often distinguishing. Interestingly, aragonite is not very stable in a geologic sense—there are no deposits of aragonite known that are older than the Carboniferous period, which ended about 300 million years ago.

Aragonite is common around the limestone deposits of the Rockies.

Augite

Large augite crystals are rare, but augite is a common component of igneous rocks.

Silicate, Ca,Na(Mg,Fe,Al)(Al,Si)$_2$O$_6$
Family: Silicates
Mohs: 5–6
Specific gravity: 3.2–3.6
Key test(s): Black, stubby crystals
Likely locale(s): Volcanic dikes

Augite is a common mineral in the pyroxene group, related to wollastonite, diopside, hedenbergite, pigeonite, and others. Augite typically forms short, stubby crystals in the monoclinic system, but they are usually too small to detect without a hand lens. Large crystals do form occasionally and can be striking and quite shiny, but they pit quickly when exposed to air. Augite has a greenish-white streak, which is a good field test.

Augite typically occurs in igneous rocks, sometimes as an essential mineral in dikes. It is common in the Rocky Mountain states of Colorado, Montana, and Idaho.

Barite

Rare blue barite crystals from Stoneham, Colorado
PHOTO COURTESY OF THE RICE NORTHWEST MUSEUM OF ROCKS AND MINERALS

Barium sulfate, BaSO$_4$
Family: Sulfates
Mohs: 3–3$\frac{1}{2}$
Specific gravity: 4.3–4.6
Key test(s): Heavy; fluorescence
Likely locale(s): Layered sedimentary rocks; veins with metal ores

There are numerous varieties of barite, including nitrobarite, radiobarite, and even fubarite. Note also that the naming convention for BaSO$_4$ may even be switching to barite in the future. Barite usually occurs as pale-yellow to brownish-white tabular crystals. Its color can range from white to dark brown, however, and even a striking blue, depending on impurities that substitute for the barium ion. Crystal structure in the field can range from slender prisms to fibrous masses. Barite has a white streak and a vitreous, glassy luster. Some barite specimens fluoresce, but after heating it to red hot and allowing it to cool, all barite appears orange under a black light.

Barite is common throughout the Rockies, especially in Montana and Colorado, and is usually dull brown or yellow. Gem-quality, facetable barite occurs in Montana in Treasure County, near Myers. Striking blue barite crystals occur in weathered shale near Stoneham, Colorado.

Beryl

Common beryl, pale green but well-formed with nice termination
PHOTO COURTESY OF THE RICE NORTHWEST MUSEUM OF ROCKS AND MINERALS

$Be_3Al_2Si_6O_{18}$
Family: Silicates
Mohs: $7\frac{1}{2}$–8
Specific gravity: 2.7–2.9
Key test(s): Hard; hexagonal form
Likely locale(s): Granite pegmatites

Like corundum, beryl comes in a variety of formulas. Gemmy bright-green beryl is an emerald, while gemmy blue beryl is aquamarine. When red or pink, beryl is morganite; other varieties include a golden beryl, which is yellow, and heliodor, which is greenish yellow. *Goshenite* is the name for colorless beryl. In all forms, beryl has a vitreous luster and a colorless streak. Crystals are hexagonal, sometimes in striking, perfectly six-sided prisms. Cleavage is not a good test, but hardness is, as beryl is harder than quartz. Common beryl is a constituent of many pegmatites in the Rockies, while gem-quality beryl typically occurs in vugs, cavities, and pockets.

Beryl is associated with pegmatites in Latah County, in the Sawtooth Mountains, and elsewhere in Idaho. Most of Wyoming's beryl comes from Fremont and Goshen Counties, but there are many other deposits. Most of Montana's beryl is associated with the Boulder Batholith. Colorado lists over one hundred beryl deposits, associated with batholiths and pegmatites mostly.

Calcite

Large calcite crystals in a cluster, from Montana (left). Calcite rhombs from Idaho (right).
PHOTOS COURTESY OF THE RICE NORTHWEST MUSEUM OF ROCKS AND MINERALS

Calcium carbonate, CaCO$_3$
Family: Carbonates
Mohs: 3
Specific gravity: 2.7
Key test(s): Rhombohedral angles; hardness
Likely locale(s): Veins

Calcite is very common and is found as crystals in pegmatites, as veins in basalts, and mixed with zeolites. It is usually yellow but can occur in a pure, clear crystal form. It has a white streak, and its luster is vitreous but sometimes very dull. One of the classic field tests for calcite is the rhombohedral crystal habit, and calcite will cleave easily along those planes. Calcite also fluoresces. Gypsum doesn't effervesce in acid. Two well-known crystal forms are dogtooth and nailhead calcite. Perfect calcite crystals display double refraction, in essence doubling the image they sit on.

Calcite is common to igneous, metamorphic, and sedimentary rocks and is widespread throughout the Rocky Mountain states.

Chalcedony

Banded chalcedony from Challis, Idaho
PHOTO COURTESY OF THE RICE NORTHWEST MUSEUM OF ROCKS AND MINERALS

Quartz, SiO$_2$
Family: Cryptocrystalline quartz
Mohs: 6$\frac{1}{2}$–7
Specific gravity: 2.6–2.64
Key test(s): No cleavage; conchoidal fracture
Likely locale(s): Quartz-rich areas

Chalcedony is one of the most common forms of quartz, and it is an all-inclusive term for a vast variety of collectible material, including agate, bloodstone, heliotrope, chrysoprase, jasper, and flint. When red or orange, chalcedony is sard, or carnelian. Other varieties include layering, making for sardonyx; green with red spots is bloodstone, or heliotrope. Rockhounds call banded or clear chalcedony agate; if there are inclusions, the sample is moss agate. Apple-green chalcedony is chrysoprase. When red, yellow, or brown, we use the term jasper, and when white or gray or dark gray, we call it flint. There is no cleavage, as there is no crystal habit, and the fracture is conchoidal. Chalcedony typically has a waxy, vitreous, or dull luster.

Chalcedony occurs in virtually all alluvial gravels to some degree. Its various forms are common throughout the Rockies.

Chrysocolla

Striking blue chrysocolla resembles turquoise.
PHOTO COURTESY OF THE RICE NORTHWEST MUSEUM OF ROCKS AND MINERALS

(Cu,Al)$_2$H$_2$Si$_2$O$_5$(OH)$_4$·nH$_2$O
Family: Silicates
Mohs: 2½–3½
Specific gravity: 1.9–2.4
Key test(s): Soft
Likely locale(s): Copper-producing regions

Chrysocolla is a pleasing blue or green material, usually banded, and softer than turquoise. It produces a streak that is slightly blue green in color, which is helpful. Its luster is vitreous to dull, and it forms in botryoidal masses or coatings rather than as crystals. Chrysocolla oxidizes from copper deposits. Its softness precludes it from being an important lapidary material, but it will take a polish.

There are numerous chrysocolla deposits throughout the copper-mining regions of the Rockies, such as the Bay Horse and Seven Devils districts of Idaho. In Wyoming search the copper-mining areas in Albany and Carbon Counties. Montana is a major copper-producing state, thus the area around Butte is noteworthy among hundreds of locales. Colorado also boasts numerous chrysocolla deposits.

Chrysotile

Fibrous chrysotile is an obvious source of asbestos fibers.
PHOTO COURTESY OF THE RICE NORTHWEST MUSEUM OF ROCKS AND MINERALS

Mg$_3$(Si$_2$O$_5$)(OH)$_4$
Family: Phyllosilicates
Mohs: $2\frac{1}{2}$–3
Specific gravity: 2.53
Key test(s): Soft, silky
Likely locale(s): Serpentine belts

Chrysotile is the only asbestos from the serpentine family; all other asbestos minerals such as tremolite and actinolite are amphiboles. Known as the "white asbestos," chrysotile is the most common form and is typically composed of white, curly fibers. It has a silky luster, and leaves a white streak. You may be able to scratch chrysotile with your fingernail, which is $2\frac{1}{2}$ on the Mohs' scale. You probably shouldn't, however, because then you'd get asbestos under your fingernail. The amphibole asbestos family forms thinner fibers and is more deadly to your lungs, but you should respect chrysotile also.

Chrysotile occurs within the Green Mountain and State Line kimberlites in Colorado. Idaho lists limited chrysotile, such as the Hailey district. Montana hosts chrysotile in the Libby area, in the Stillwater Complex, and it is concentrated in Madison County. Wyoming hosts chrysotile among scattered asbestos deposits in Natrona County and Converse County.

Corundum

Corundum in the form of very low-grade ruby (left). Sapphires (right) from Spokane Bar, Montana.

Aluminum oxide, Al_2O_3
Family: Oxides
Mohs: 9
Specific gravity: 3.9–4.1
Key test(s): Only thing harder is a diamond
Likely locale(s): Alluvial deposits

North American corundum comes in three main varieties: emery, which is black or dark gray and historically used for abrasives; sapphire, which can be blue, yellow, purple, or green; and rubies, which are red. Each variety has a hardness of 9 on the Mohs' scale, so all forms of corundum scratch only by diamond. Luster ranges from vitreous all the way to adamantine, especially in true, gem-quality rubies and sapphires. The hexagonal crystal habit is not a good field indicator, mostly because it is rare to find crystals big enough to inspect. The hardness test is easier in any case.

Wyoming's Fremont County hosts a handful of corundum deposits; it is sparse elsewhere in the state. Idaho is similarly limited, with Burgdorf, Rock Flat, the Lochsa River, and Clearwater County prominent. Montana and Colorado are more noteworthy, with Montana topping the list and containing dozens of deposits.

Epidote

Well-formed epidote crystals (left). Epidote bar from Fix Ridge, near Genesee, Idaho (right).
PHOTOS COURTESY OF THE RICE NORTHWEST MUSEUM OF ROCKS AND MINERALS

$Ca_2Al_2Fe^{3+}OSiO_4Si_2O_7(OH)$
Family: Silicates
Mohs: 6
Specific gravity: 3.3–3.6
Key test(s): Color
Likely locale(s): Metamorphic rocks

Epidote is a common rock-forming mineral, with a characteristic green color, although it can occur as yellow green or even greenish black. Epidote has a colorless to gray streak. Crystals are common in the monoclinic system, forming long, slender crystals or as massively tabular encrustations. Epidote occurs in many situations: in epidote schist; in granite pegmatites; in basalt cavities; along with andesite; in greenstones; and in regional metamorphic rocks.

Epidote is common across the Rocky Mountain states. Wyoming is probably the least represented, but Idaho, Montana, and Colorado have hundreds of known epidote locales.

Feldspars–Alkali–Microcline

Giant microcline crystals with black smoky quartz from the Sawtooth Mountains in Idaho
PHOTO COURTESY OF THE RICE NORTHWEST MUSEUM OF ROCKS AND MINERALS

KAl AlSi$_3$O$_8$
Family: Tectosilicates
Mohs: 6–6½
Specific gravity: 2.5–2.6
Key test(s): Hardness
Likely locale(s): Common rock-forming mineral

Alkali feldspars, sometimes known as potassium feldspars or potash feldspars, are usually white, light gray, or light pink. One exception is amazonite, which appears light blue or green. These varieties of microcline have a vitreous luster and leave a white streak. Feldspars are key rock-forming minerals for igneous and metamorphic rocks and make up more than 50 percent of the earth's crust.

In Idaho there are scattered locales across the state, with prized specimens coming from the Sawtooths. Montana also has numerous deposits, with good crystals in the Boulder Batholith, at Lemhi Pass, and elsewhere. Microcline occurs in numerous pegmatites in Wyoming, especially in Goshen County and Albany County. Colorado has hundreds of microcline locales throughout the western and central regions.

Feldspars—Alkali—Microcline var. Amazonite

Beautiful blue amazonite crystals with black smoky quartz from Lake George, Colorado
PHOTO COURTESY OF THE RICE NORTHWEST MUSEUM OF ROCKS AND MINERALS

KAl AlSi$_3$O$_8$
Family: Tectosilicates
Mohs: 6–6$\frac{1}{2}$
Specific gravity: 2.5–2.6
Key test(s): Hardness
Likely locale(s): Pegmatites

Amazonite shares all the identification properties of microcline, but amazonite ranges from pale, light green to a deep, greenish blue. Prized specimens contain more color, as most amazonite is not as vivid as the museum-grade pieces. Amazonite occurs in large vugs and cavities within pegmatites, combining with smoky quartz and phenakite.

Colorado is the only state in the Rockies to produce amazonite. Pegmatites on private claims in the Lake George area yield very high-quality specimens.

Feldspars—Plagioclase—Orthoclase

Orthoclase crystals in matrix (left) and single weathered crystal (right) from Mosquito Pass, Colorado
PHOTOS COURTESY OF THE RICE NORTHWEST MUSEUM OF ROCKS AND MINERALS

Orthoclase: $(KAl, NaAl, or CaAl_2)Si_3O_8$
Family: Tectosilicates
Mohs: $6–6\frac{1}{2}$
Specific gravity: 2.6–2.8
Key test(s): Hardness
Likely locale(s): Common rock-forming mineral

Plagioclase feldspars rank by how much anorthite is present, ranging from zero to 100 percent, from pure albite to pure anorthite. These common rock-forming minerals have a vitreous luster and a white streak. The color is variable but is typically dull white or light gray in appearance, with light tinting common. Feldspars show good cleavage in two directions at ninety degrees, and good twinning. Adularia is a transparent. Sunstones are a form of oligoclase feldspar. Labradorite is another gem variety of orthoclase feldspar.

Orthoclase isn't noteworthy throughout most of the Rocky Mountain states. Colorado's Pitkin and El Paso Counties offer collectible orthoclase specimens.

Fluorite

Light-blue fluorite crystals from Challis, Idaho
PHOTO COURTESY OF THE RICE NORTHWEST MUSEUM OF ROCKS AND MINERALS

Calcium fluoride, CaF$_2$
Family: Halides
Mohs: 4
Specific gravity: 3.0–3.2
Key test(s): Pastel purple, pink, green color; cubic crystal habit
Likely locale(s): Found in hydrothermal, sedimentary, igneous, and volcanic deposits

MINERALS

Fluorite comes in a variety of colors such as purple, green, and pink, and banding is common. It leaves a white streak, as do many minerals, so a better test is the hardness test or the square crystals. Fluorite crystals are unique in that they cleave perfectly in four directions. Fluorite has a vitreous luster and is softer than quartz but harder than calcite. Another good field test is to check for fluorescence. Fluorite veins are often associated with economic ore deposits, such as galena and sphalerite.

Idaho's famed Challis area offers decent fluorite samples, as do Meyers Cove and other scattered locales. Montana and Wyoming also offer limited fluorite collecting. The best state by far in the Rockies for collecting gem-quality fluorite is Colorado, especially in Jefferson, Boulder, and Teller Counties.

Fulgurite

This fulgurite from Boulder County, Colorado, formed when lightning fused quartz-rich sands.
PHOTO COURTESY OF THE IMAGE ARCHIVES, DENVER MUSEUM OF NATURE AND SCIENCE

Quartz, SiO$_2$
Family: Quartz
Mohs: 7
Specific gravity: 2.65
Key test(s): Tubelike fused glass; rough exterior
Likely locale(s): Sandy deserts

Sand fulgurites are the result of lightning strikes hitting quartz-rich sand deposits. (Rock fulgurites result from lighting striking rocks, usually at the tops of mountains, and are less collectible.) The intense heat of the strike instantly fuses the quartz in the sand, usually in a tube that is rough on the outside and smooth, sometimes bubbly, on the inside. The resulting color is based on the melted material and is usually gray, light gray, or tan. Fulgurites can extend several feet into the ground, usually branching or tapering away eventually.

Look for fulgurites in any flat, open sandy desert or dune area where thunderstorms are common. Wyoming's Red Desert would be a good place to look, or in Boulder County, Colorado.

Garnet

Rough garnet from Emerald Creek, Idaho

$X_3Y_2(SiO_4)_3$ Almandine
Family: Metal silicates
Mohs: $6\frac{1}{2}-7\frac{1}{2}$
Specific gravity: 3.6–4.3
Key test(s): Hackly fracture, hardness
Likely locale(s): Schist; black sands

Increasing metamorphism $\longrightarrow$

Chlorite	Biotite	Garnet	Staurolite	Kyanite	Sillimanite

Taken as a family, garnet crystals show a vitreous luster, have no streak to speak of, and do not cleave. Garnet is harder than apatite, does not fluoresce like zircon, and has higher specific gravity than tourmaline, plus it is usually associated with schist. That rule isn't fixed, however, as some gem garnets are associated with granitic pegmatites. Gem-quality garnet is somewhat rare, and adding further value is a tendency for impurities to line up and create four-sided or six-sided stars.

Garnets are common across the Rockies. Wyoming has the fewest locales, mostly in Goshen County. Colorado and Montana have dozens of locales. Ruby Reservoir in Montana is an easy collecting spot. Some streams in northern Idaho run red with millions of tiny garnets plentiful enough to supply sandpaper companies with raw product.

Geodes

Rough geode from Idaho's famed Bruneau Canyon (left) and brecciated double egg from Marsing, Idaho (right)
PHOTO COURTESY OF THE RICE NORTHWEST MUSEUM OF ROCKS AND MINERALS

Quartz, SiO$_2$
Family: Silicates
Mohs: Varies
Specific gravity: Varies
Key test(s): Round shape
Likely locale(s): Rhyolite lava beds

Geodes are round, quartz-filled structures that range greatly in size and shape. They differ from nodule and concretions in their formation; nodules and concretions occur in sedimentary rocks. Some geodes are almost perfectly round, about the size of a small fist, and when completely full of quartz, agate, or chalcedony, are usually called thunder eggs. Some geodes are mis-shaped and hollow.

Bruneau Canyon in Idaho produces some fine geodes. The Rabbit Springs area, on the Idaho-Nevada border near Jackpot, Nevada, produces mostly inferior, hollow geodes. In Colorado, reports point to Wolf Creek Pass, Pueblo Reservoir, and several other places.

Gypsum

Spar variety of gypsum (left) and Ram's Horn variety (right)
PHOTOS COURTESY OF THE RICE NORTHWEST MUSEUM OF ROCKS AND MINERALS

Calcium sulfate, $CaSo_4 \cdot 2H_2O$
Family: Sulfates
Mohs: $1\frac{1}{2}$–2
Specific gravity: 2.3–2.4
Key test(s): Softer than calcite
Likely locale(s): Hydrothermal basins

Gypsum is usually white in hand specimens but sometimes appears colorless in pure crystals. It leaves a white streak. Crystals are monoclinic; a rhombic pattern is common, as is massive, fibrous "satin spar" form. The Ram's Horn variety is not common, but it is striking and worth noting. Cleavage is perfect in one direction. Gypsum is softer than calcite, which is an easy field test. It is common in hydrothermal replacement deposits and in sedimentary basins, resulting in giant strip-mining operations for use in sheet rock or as fertilizer. One form of gypsum, selenite, was discovered in a Mexican cave in the form of 30-foot crystals.

There are numerous gypsum deposits across the Rockies, including southeast Idaho, along the border with Wyoming; and throughout Wyoming, Montana, and Colorado.

MINERALS

Halite

Cubic halite showing excellent cleavage and vitreous luster
PHOTO COURTESY OF THE RICE NORTHWEST MUSEUM OF ROCKS AND MINERALS

Sodium chloride, NaCl
Family: Chlorides
Mohs: 2–2½
Specific gravity: 2.16
Key test(s): Taste
Likely locale(s): Evaporite deposits; sedimentary

Halite, or rock salt, is usually clear but is sometimes tinted pink or gray. It leaves a white streak. Halite is strongly isometric, and when pure, features well-formed crystals in tiny cubes. It has a vitreous luster and perfect cleavage in three directions. Halite is similar to cryolite but harder, and it has a salty taste. Halite typically occurs in large evaporite deposits and in large dome-like underground deposits. It mixes readily with other ions, and there are perhaps twenty more halide minerals, such as sulphohalite and polyhalite.

In Wyoming, Albany and Sweetwater Counties have yielded halite, with some deposits known to Oregon Trail pioneers. Colorado also has numerous halite deposits, especially in Rio Blanco County.

Hornblende

These coarse, black hornblende crystals are larger than normal.

$(Ca,Na,K)2-3(Mg,Fe_2+,Fe_3+,Al)_5(SiAl)_8O_{22}(OH)_2$
Family: Inosilicates
Mohs: 5–6
Specific gravity: 3.0–3.4
Key test(s): Cleavage angles
Likely locale(s): Common rock-forming mineral

Hornblende is a common rock-forming mineral, usually black but sometimes green or brown. It is actually the name given to a series of amphiboles, all differing by their amounts of iron, magnesium, calcium, sodium, aluminum, and potassium. Hornblende leaves a colorless streak, and crystals are monoclinic, usually short, and prismatic. Cleavage is perfect in two directions. It can be confused with schorl, which is black tourmaline, but schorl does not cleave like hornblende. Diamond-shaped cleavage angles and a greenish-black color are the best field indicators.

Hornblende is rarely collectible as large crystals, but it is common across the mining districts of all Rocky Mountain states.

Ilvaite

Large blades and stubby crystals of ilvaite from South Mountain, Idaho
PHOTO COURTESY OF THE RICE NORTHWEST MUSEUM OF ROCKS AND MINERALS

$CaFe_2+2Fe_3+(Si_2O_7)O(OH)$
Family: Silicates
Mohs: $5\frac{1}{2}$–6
Specific gravity: 4.05
Key test(s): Black streak
Likely locale(s): Contact metamorphic zone

Ilvaite takes its name from *Ilva*, the old Latin pronunciation for Elba Island, where Napoleon spent his exile. That is the type locality for the mineral. Ilvaite occurs as shiny black stubby crystals with good cleavage and brittle tenacity. It generally leaves a black streak but may show subtle green or brown powder. Ilvaite also occurs as black, dull masses, making it harder to identify.

Idaho's South Mountain supplies the only fine ilvaite specimens in the Rocky Mountain states.

Jasper

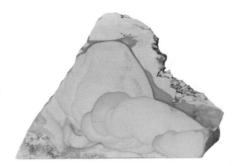

Polished picture jasper from Willow Creek, Idaho

Quartz, SiO$_2$
Family: Cryptocrystalline quartz
Mohs: 7
Specific gravity: 2.65
Key test(s): Conchoidal fracture
Likely locale(s): Basalt flows

Jasper is another common form of quartz, derived from silica-rich ground solutions typically circulating through basalt, which technically makes it a form of chalcedony. Jasper comes in many shades and colors but is usually red, tan, or yellow and sometimes green. It has no crystal habit and leaves a white streak. Gem-quality jasper is hard and shiny, while common jasper is typically porous and will not polish. Jasper occurs in most regions where basalt is common, but not always.

Jasper is common in the Rockies, with significant collecting opportunities in every state. Bruneau picture jasper from Idaho commands top prices, as does Willow Creek picture jasper from Gem County, north of Boise. Beware of multiple mining claims in the most popular collecting areas.

Kyanite

Kyanite usually occurs as dark-blue blades.

$Al_2(SiO_4)O$
Family: Aluminosilicates
Mohs: $5\frac{1}{2}$–7
Specific gravity: 3.5–3.7
Key test(s): Blue, bladed crystals; splinters easily
Likely locale(s): Metamorphic terrain; pegmatites

Increasing metamorphism ⟶

Chlorite	Biotite	Garnet	Staurolite	Kyanite	Sillimanite

Kyanite is an indicator mineral that reveals an area has undergone intense regional metamorphism. It is typically blue, sometimes deep blue, and usually found in long, columnar crystals that appear fibrous or bladed. Crystals are often brittle. Kyanite is a great representative of an obscure mineralogical test known as anisotropism. Most minerals do not care how you measure their hardness, but kyanite is an exception. When you measure the hardness of kyanite, you will measure it at 7 if you go lengthwise along the blade, but only $5\frac{1}{2}$ if you go across the short axis.

Kyanite is common in northern Idaho, especially in Clearwater, Latah, and Shoshone Counties; Freezeout Ridge is a popular locale. Western Montana has numerous kyanite deposits associated with pegmatites and corundum and sapphire deposits. Wyoming's Laramie Range contains kyanite deposits. In Colorado the Robinson Gulch pegmatite, the Famous Lodge pegmatite, and the Sedalia Mine all host kyanite.

Mica Group—Biotite

Biotite sheets and flakes are usually quite dark.
PHOTO COURTESY OF THE RICE NORTHWEST MUSEUM OF ROCKS AND MINERALS

Biotite: $K(Mg,Fe)_3(Al,Fe)Si_3O_{10}(OH,F)_2$
Family: Phyllosilicates
Mohs: 2–3
Specific gravity: 2.7–3.0
Key test(s): Platy cleavage
Likely locale(s): Schists

Increasing metamorphism ⟶

Chlorite	Biotite	Garnet	Staurolite	Kyanite	Sillimanite

MINERALS

The mica group is composed of several similar sheetlike minerals, including muscovite, biotite, phlogopite, lepidolite, chlorite, and others. Of the two most common varieties, muscovite is white or colorless; biotite is much darker. In fact, mica experts recommend that we no longer use the name biotite for a specific chemical formula; instead, they suggest you use "biotite" as a field term for all dark micas you encounter. Well-formed micas occur in granitic pegmatites, a historic source of muscovite, while phlogopite resides in marble and hornfels.

Dark-black or brownish-black biotite micas are common throughout the Rockies, usually in schists and pegmatites.

Mica Group—Chlorite

This orthoclase feldspar is dusted with a coating of light-green chlorite.
PHOTO COURTESY OF THE RICE NORTHWEST MUSEUM OF ROCKS AND MINERALS

$(Mg,Fe)_3(Si,Al)_4O_{10}(OH)_2 \cdot (Mg,Fe)_3(OH)_6$
Family: Phyllosilicates
Mohs: 2–2½
Specific gravity: 2.6–3.3
Key test(s): Green, soft; platy cleavage
Likely locale(s): Schists; hydrothermal ore deposits

Increasing metamorphism ⟶

Chlorite	Biotite	Garnet	Staurolite	Kyanite	Sillimanite

Like biotite, the term chlorite no longer refers to a single mineral. Instead, it is the name for a group of related mica minerals, including clinochlore, chamosite, cookeite, and several others. However, field workers still use the term "chlorite" to refer to green coatings or stains on other minerals. Chlorite represents the lowest grade of metamorphism; it is a key constituent of greenschists and greenstones, for example. Chlorite has a white streak and is generally not collectible except for the rare forms.

Chlorite is common throughout the low-grade metamorphic rocks of the Rocky Mountains.

Mica Group—Muscovite

Muscovite sometimes occurs in large sheets called "books."
PHOTO COURTESY OF THE RICE NORTHWEST MUSEUM OF ROCKS AND MINERALS

$KAl_2(AlSi_3O_{10})(F,OH)_2$
Family: Phyllosilicates
Mohs: 2–2½ parallel to main cleavage, 4 at right angle
Specific gravity: 2.8–3.0
Key test(s): Platy cleavage; pearly luster
Likely locale(s): Schists

Increasing metamorphism ⟶

Chlorite	Biotite & Muscovite	Garnet	Staurolite	Kyanite	Sillimanite

Muscovite is the most common mica seen in the field and is usually white or colorless; biotite is much darker. Chlorite is greenish, while lepidolite is pink or light purple. Muscovite has a vitreous, pearly luster, leaves a colorless streak, and perhaps most characteristic, its cleavage is perfect in one direction, resulting sometimes in great sheets called "books." Because muscovite flakes can look yellow and shiny, they are sometimes mistaken for fool's gold, but a knifepoint will fracture mica flakes, while flattening or cutting gold. If large enough, mica show a hexagonal pattern. Muscovite bends the easiest of the mica group.

Muscovite is common across the Rockies, especially in pegmatites. In Montana, muscovite is common north and west of Butte, and in Colorado it is associated with pegmatites among the granite peaks.

Olivine

Olivine-rich basalt in the ground at Dunraven Pass, inside Yellowstone National Park

$(Mg,Fe_2+)2SiO_4$
Family: Nesosilicates
Mohs: $6\frac{1}{2}$–7
Specific gravity: 3.5–4.0
Key test(s): White streak
Likely locale(s): Found with metamorphic rocks and in hydrothermal replacement deposits

Olivine is usually dark green, with a vitreous luster and colorless streak. When found in crystalline form, such as the semiprecious gemstone peridot, olivine will cleave in two directions, but it is more commonly associated with olivine-rich basalt. *Olivine* is the name for a series of minerals, with forsterite on the magnesium-rich end of the series and by far the most common, and fayalite on the iron-rich end. Olivine is actually a common rock-forming mineral, comprising much of the earth's mantle, and there are beaches in Hawaii that have turned green from the presence of so much olivine.

Olivine is mostly associated with asbestos deposits in Idaho. It is a major component of the Stillwater Complex in Montana and occurs in scattered deposits elsewhere in the state. Olivine occurs in a few places in Wyoming, with an olivine-rich basalt prominent at Yellowstone National Park. Colorado contains numerous olivine locales, especially in the mountainous mining districts.

Onyx

Sliced and polished onyx
PHOTO COURTESY OF THE RICE NORTHWEST MUSEUM OF ROCKS AND MINERALS

Quartz, SiO$_2$
Family: Quartz
Mohs: 6–7
Specific gravity: 2.6–2.7
Key test(s): Patterns
Likely locale(s): Silica-rich environments

Onyx is yet another form of the chalcedony family of quartz. Look for uniform banding, creating striking patterns of red, white, black, or brown. If the bands contain sard, the result is termed sardonyx. The most common color is black, but rockhounds often label as onyx material formed in white carbonates such as limestone, marble, and calcite. Mineralogists restrict onyx to a variety of agate. The luster is vitreous to silky, streak is white, and onyx is usually slightly translucent.

Onyx is limited in the Rocky Mountain states. There are reports of onyx deposits in Grand and Saguache Counties in Colorado. Montana lists the Montana Onyx Mine in Madison County's Ruby Range. The Hartville Onyx Mine is in Platte County, Wyoming.

Opal

Common opal from within basalt flows

Quartz, SiO$_2$·nH$_2$O
Family: Cryptocrystalline quartz
Mohs: 5$\frac{1}{2}$–6$\frac{1}{2}$
Specific gravity: 2.0–2.2
Key test(s): Softness, lighter than other quartz
Likely locale(s): Quartz-rich areas

Opal is another form of quartz, but this time with considerable water present. There are numerous varieties of common opal, with mindat.org showing at least 155 names. Common opal, sometimes referred to as hyalite, has a distinctive vitreous, pearly luster. It is usually white, but colors typically include yellow, brown, green, blue, or gray. Its streak is white. Opal is not crystalline; there is no habit or cleavage, and it has a conchoidal fracture. Opal is softer than agate, jasper, and other forms of quartz. Opal frequently forms in cavities, fractures, and air bubbles in basalt, where it sometimes weathers out and remains as small, round pea-size blebs. Wood and shells often become "opalized" through replacement. Opal will fluoresce, which also sets it apart from jasper and quartz.

In Idaho opal shows up as a nuisance rock among diatomite deposits in Ada and Elmore Counties. In central Wyoming geologist W. Dan Hausel discovered enormous opal deposits, mostly common opal. Montana offers scattered minor opal deposits across the state. Colorado has dozens of known opal outcrops and deposits, especially in Chaffee County.

Orpiment and Realgar

Red orpiment (left) and yellow realgar (right)
PHOTOS COURTESY OF THE RICE NORTHWEST MUSEUM OF ROCKS AND MINERALS

As$_2$O$_3$ and As$_4$S$_4$
Family: Arsenides
Mohs: 1$\frac{1}{2}$–2
Specific gravity: 3.5–3.6
Key test(s): Orpiment is bright yellow; realgar is deep red
Likely locale(s): Hot springs and fumaroles

These two arsenic minerals usually show up with each other; orpiment is an oxide of arsenic, and realgar is a sulfide. Orpiment has a bright-yellow streak; ancient artists and painters used orpiment as a yellow pigment. Realgar is deep orange-red. Orpiment is often a by-product of weathered, altered realgar, where the sulfur leaches away, leaving an oxide. Likely collecting locales include hot springs and hydrothermal vein systems. Orpiment rarely occurs as a crystal and is much more common as a fibrous, velvety coating, sometimes in the botryoidal habit. Realgar often displays as small, granular crystals, so larger specimens are highly sought.

Orpiment is associated with mercury at the Hermes Mine in Valley County, Idaho; realgar is common in the Black Pine district. Montana hosts orpiment and realgar in the Philipsburg district and in the South Moccasin Mountains. In Wyoming the Encampment and New Rambler districts to the southeast and the South Pass district in the west both contain orpiment and realgar specimens. In Colorado the San Luis Hills in Conejos County contain arsenides, as does the La Plata district.

Phenakite

Water-clear rhombohedral phenakite specimen with excellent crystal faces
PHOTO COURTESY OF THE RICE NORTHWEST MUSEUM OF ROCKS AND MINERALS

Be_2SiO_4
Family: Nesosilicates
Mohs: $7\frac{1}{2}$–8
Specific gravity: 2.96
Key test(s): Hardness
Likely locale(s): Pegmatites

Phenakite occurs in a wide variety of disguises. It can be tabular, it can be columnar, and it can even occur in granular sheets. It can be colorless, white, gray, or even pale pink. It has a striking, colorless crystal that is occasionally gemmy enough to facet. Phenakite is harder than $7\frac{1}{2}$ and thus will scratch quartz. Phenakite and beryl are equally hard, but beryl is hexagonal and usually forms longer crystals.

Phenakite occurs in both Idaho and Colorado. In Idaho the Sawtooth Batholith in Boise County and a pair of locales in Elmore County dominate the literature. Colorado contains striking rhombohedral phenakite, in association with amazonite in the pegmatites near Lake George in Park County. More elongated phenakite crystals originate in Colorado's Mount Antero region.

Quartz

Smoky quartz from Bald Mountain, Montana (right), and straight, clear quartz (left)
SMOKY QUARTZ PHOTO COURTESY OF THE RICE NORTHWEST MUSEUM OF ROCKS AND MINERALS

Crystalline quartz, SiO_2
Family: Silicates
Mohs: 7
Specific gravity: 2.65
Key test(s): Hardness; glossy black color
Likely locale(s): Pegmatites

Smoky quartz is a semiprecious gem variety of common crystalline quartz, so it shares the same diagnostic properties. The hardness is 7, there is a white streak if you can produce it, and crystals are hexagonal, with no cleavage. Collectors prize the specimens with a characteristic pointed, pyramidal termination. The black coloring appears to be the result of free silicon atoms that underwent radiation from the natural background radiation common to granite. Gem-quality smoky quartz crystals are associated with granite pegmatites, often accompanied by feldspar.

Because there are so many pegmatites in the Rocky Mountain states, smoky quartz is abundant in the region. One good collecting locale is off US 12 along the Idaho-Montana border at Lolo Pass. Idaho's Sawtooth Mountains contain multiple smoky quartz locales, and smoky quartz is associated with striking blue amazonite crystals near Lake George in Colorado.

Rhodochrosite

The Alma Rose, one of the finest rhodochrosite specimens in the world, is from the Sweet Home Mine in Colorado.
PHOTO COURTESY OF THE RICE NORTHWEST MUSEUM OF ROCKS AND MINERALS

Manganese carbonate, $MnCO_3$

Family: Carbonates

Mohs: $3\frac{1}{2}$–4

Specific gravity: 3.7

Key test(s): Deep pink to rose red; white streak, vitreous luster

Likely locale(s): Vein mineral in silver districts

Rhodochrosite is a lovely rosy red in its pure form, but impurities typically temper that vivid color down to pink, cinnamon, or even yellow. Its rhombohedral crystals are rare and highly collectible, but most collectors encounter field deposits as ribbons or banded minerals. These specimens will often take a polish and produce striking red-pink, banded lapidary creations. As crystals, rhodochrosite is soft and difficult to fashion into jewelry, and it has perfect cleavage, another strike against it as far as faceting.

Alma, Colorado, boasts some of the best rhodochrosite in the world, including the Alma Rose, a deep-red, beautiful crystal. Wyoming has a single showing in Fremont County in the literature, and Idaho has about a dozen locales, mostly in the Panhandle area and the Sawtooth Mountains. Montana has about one hundred locales, with nice specimens from the Emma Mine in Butte County.

Rhodonite

Light-pink color is characteristic of rhodonite, as are the dark-manganese "roads" through the pink material.
PHOTO COURTESY OF THE RICE NORTHWEST MUSEUM OF ROCKS AND MINERALS

Manganese silicate, MnSiO$_3$
Family: Metal silicates
Mohs: 5½–6½
Specific gravity: 3.6–3.8
Key test(s): Pink (when fresh), turns brown
Likely locale(s): Metamorphic rocks; hydrothermal replacement deposits

Rhodonite is usually easy to identify because it is rose pink in the field and is one of the few minerals found in that color. Some rhodonite samples are red, brown, or yellow, depending on the calcium, iron, or magnesium impurities present with the manganese, so the color test is not foolproof. Most rhodonite samples are massive and pink, with dendrites of psilomelane that look like tiny roads. You can use that "road-like" black tracking as a memory aid—roads in rhodonite.

There are rhodonite deposits across Montana's mining districts, and especially around Butte. Idaho's literature is much more limited, listing a single showing in the Edwardsburg district of Valley County. Wyoming is apparently barren of rhodonite, but Colorado has dozens of deposits, especially in the San Juan Mountains.

Selenite

Splendid selenite crystals from Cave of Swords in Chihuahua, Mexico
PHOTO COURTESY OF THE RICE NORTHWEST MUSEUM OF ROCKS AND MINERALS

$CaSO_4 \cdot 2H_2O$
Family: Sulfates
Mohs: 2
Specific gravity: 2.3
Key test(s): Water-clear crystals
Likely locale(s): With gypsum

Selenite is a form of gypsum and is a general term for satin spar, desert rose, gypsum flower, and selenite crystals. It does not contain the element selenium. Rockhounds typically reserve the name *selenite* for superior, transparent gypsum crystals found either individually or as masses. Selenite comes in a variety of colors, including brownish green, brownish yellow, gray white, and other light tints. Primarily, however, think of the most water-clear tabular and columnar crystals as being selenite. Selenite has a white streak and a pearly luster.

Montana offers selenite near Butte and at the White Cloud Mine in Ravalli County. Idaho is limited in selenite showings, with a single mention in the Lava Creek district in Butte County. Wyoming contains selenite specimens at the Grand Rapids Gypsum Mine; near Thermopolis; in Hanna Basin; and elsewhere. Colorado has multiple locales, including in San Juan County and El Paso County.

Sillimanite

Various rounded sillimanite pebbles. This mineral is difficult to locate as pure crystals.
PHOTO COURTESY OF THE RICE NORTHWEST MUSEUM OF ROCKS AND MINERALS

Al$_2$SiO$_5$

Family: Nesosilicates

Mohs: 7

Specific gravity: 3.2

Key test(s): Silky, fibrous

Likely locale(s): Severely metamorphosed aluminum-rich sedimentary rocks

Increasing metamorphism ——→

Chlorite	Biotite	Garnet	Staurolite	Kyanite	Sillimanite

Sillimanite is an indicator mineral for rocks that have undergone metamorphism at the highest temperature. In other words, rocks with sillimanite got hotter than rocks with kyanite—even though the two minerals share the same formula (along with andalusite). Sillimanite crystals are usually rare and small; it is more common to find rocks with lots of sillimanite present, and they sometimes appear to shimmer. This is due to the silky luster, the fibrous mass of crystals, and the wavy nature of their alignment.

Colorado contains almost one hundred sillimanite locales in its metamorphic terrains, including Fremont County. Sillimanite pebbles are easy to find in the lower Clearwater River gravels of Idaho. Clearwater, Latah, Lemhi, and Shoshone Counties in Idaho all contain sillimanite locales. In Wyoming sillimanite is associated with the Laramie Range and the Anderson pegmatite. In Montana's Ruby range in Madison County, sillimanite is associated with kyanite.

Staurolite

Staurolite schist from Bathtub Mountain, Idaho (left), and staurolite twins (right)
STAUROLITE TWINS PHOTO COURTESY OF THE RICE NORTHWEST MUSEUM OF ROCKS AND MINERALS

$Fe_2Al_9Si_4O_{22}(OH)_2$
Family: Nesosilicates
Mohs: 7–7½
Specific gravity: 3.7–3.8
Key test(s): Hardness, twinning
Likely locale(s): Highly metamorphosed schists

Increasing metamorphism ⟶

Chlorite	Biotite	Garnet	Staurolite	Kyanite	Sillimanite

Staurolite, which mainly occurs in staurolite schist, is an indicator mineral for intense metamorphism. A staurolite schist denotes a regionally metamorphosed rock between garnet and kyanite grade. Staurolite crystals are usually brown, with a vitreous luster and a white streak. The crystal habit is monoclinic; twinning, such as right-angle "fairy" crosses, is common. Cleavage is poor, in lengthwise direction.

The most popular Idaho collecting locales for staurolite are Bathtub Mountain and Carpenter Creek, in Shoshone and Benewah Counties, where garnet schist, staurolite schist, and kyanite schist all crop out. Crystal Cross Mountain in Montana contains kyanite and staurolite. Fremont County and Larimer County in Colorado both contain staurolite.

Talc

Talc is so soft, it is easy to carve with your fingernail.
PHOTO COURTESY OF THE RICE NORTHWEST MUSEUM OF ROCKS AND MINERALS

$Mg_3Si_4O_{10}(OH)_2$
Family: Magnesium silicates
Mohs: 1
Specific gravity: 2.7–2.8
Key test(s): Soft
Likely locale(s): Metamorphic terrain

Talc is easy to test in the field because you can scratch it with your fingernail—it represents 1 on the Mohs' hardness scale. It is the main component of most soapstones and usually occurs as a mix of magnesium silicates, such as tremolite or magnesite. Beware of the tendency for asbestos to accompany talc. Color varies, depending on associated minerals present, and can range from white to bright green. The streak is white. Talc has a greasy feel and shows a pearly luster. Crystals are extremely rare. Look for talc in metamorphic regions with ultramafic rocks such as dunite or peridotite.

In Idaho talc is associated with asbestos mining in Clearwater and Idaho Counties. In Montana Beaverhead County and Madison County dominate talc deposit listings, along with the Stillwater Complex. Rockhounds collect soapstone near Alder, Montana. In Wyoming talc is associated with some jade deposits in Fremont County and asbestos mines in Halleck Canyon. Colorado has multiple talc deposits in its mountains.

Tourmaline—Schorl

Black tourmaline crystals form a zone in this granite pegmatite from Larimer County, Colorado (left). Large, chunky crystal of schorl (right).
TOURMALINE PHOTO COURTESY OF THE IMAGE ARCHIVES, DENVER MUSEUM OF NATURE AND SCIENCE

$Na(Mg,Fe)_3Al_6(BO_3)_3(Si_6O_{18})(OH,F)_4$

Family: Cyclosilicates
Mohs: $7-7\frac{1}{2}$
Specific gravity: 3.0–3.3
Key test(s): Harder than apatite
Likely locale(s): Pegmatites

Tourmaline comes in a variety of forms, with the most common being black schorl. That is by far the most common tourmaline species, but semiprecious gem varieties include elbaite, indicolite, rubellite, and dravite. Tourmaline has a vitreous luster, leaves a white streak, and forms a hexagonal crystal with no cleavage. Hornblende is also black, but schorl has a triangular cross section.

In Colorado, Gunnison County, Fremont County, and Clear Creek County all yield varieties of tourmaline. In Wyoming try pegmatites in the South Pass area, in Albany County, and elsewhere. In Montana tourmaline is abundant. The Boulder Batholith in Jefferson County yields excellent tourmaline specimens. Idaho's Clearwater and Latah Counties both feature collectible tourmaline, including the Muscovite Mine at Mica Mountain.

Variscite

Variscite specimen from the Turquoise Mountain Mine in Fremont County, Idaho
PHOTO COURTESY OF THE RICE NORTHWEST MUSEUM OF ROCKS AND MINERALS

$AlPO_4 \cdot 2H_2O$
Family: Phosphates
Mohs: $4\frac{1}{2}$
Specific gravity: 2.6
Key test(s): Greenish; veined
Likely locale(s): Phosphate regions

Variscite is a rare secondary mineral formed when phosphate-rich groundwater leaches through aluminum-rich rocks. It usually forms as attractive greenish-blue masses or nodules, frequently shot through with white, brown, or black veins. It is lighter than malachite and greener than turquoise, two minerals with which it can be confused. Variscite has a white streak, a vitreous-to-waxy luster, and takes a nice polish. There are many known varieties of this mineral, depending on the ratio of aluminum and iron, the presence of arsenic, and the amount of other elements available.

Variscite occurs at Two Top Creek in Fremont County, Idaho. In Colorado search the Columbine district in Routt County.

Zeolite Group: Heulandite

Nice pink heulandite specimen from the Rat's Nest Mine, near Challis, Idaho
PHOTO COURTESY OF THE RICE NORTHWEST MUSEUM OF ROCKS AND MINERALS

$(Ca,Na)(Al_2Si_7O_{18})\cdot6H_2O$
Family: Zeolites
Mohs: $3-3\frac{1}{2}$
Specific gravity: 2.2
Key test(s): Pink (when fresh)
Likely locale(s): Basalt flows and hornfels

Heulandite is the name for a group of zeolites known for a distinctive pink-colored form, but often crystals are white or lack color completely. The streak is white. Crystals are monoclinic and sometimes display a unique coffin shape. Luster can be pearly or vitreous. Heulandite is usually found with other zeolites in veins or pockets within basalt and andesite but can also occur in schists.

The Challis area of central Idaho boasts excellent heulandite specimens; elsewhere in the state, check along US 95 in Idaho County and in the Sawtooths. In Colorado search the San Juan Mountains and Jefferson County, among others. Wyoming offers only limited zeolite hunting; Montana also mostly offers micro-specimens.

Zeolite Group: Mordenite

Superb mordenite, said to be the world's finest, from the Rat's Nest Mine near Challis, Idaho
PHOTO COURTESY OF THE RICE NORTHWEST MUSEUM OF ROCKS AND MINERALS

$(Ca, Na_2, K_2)Al_2Si_{10}O_{24} \cdot 7H_2O$
Family: Zeolites
Mohs: 3–4
Specific gravity: 2.1
Key test(s): Fibrous
Likely locale(s): Veins and cavities in basalt flows

Mordenite forms sprays of delicate needles, which makes it one of the easier zeolites to identify. It is usually white, although it can discolor quickly if it begins to absorb moisture. It streaks white or colorless and usually forms as needlelike crystals but can also form velvety coatings. It usually occurs in the cavities and vugs of volcanic rocks.

Challis, Idaho, supplies some of the best mordenite specimens found anywhere. The road cuts along US 95 near Pinehurst, Idaho, are also excellent. In Montana try the Frying Pan Basin, and since you can only collect outside of Yellowstone National Park, try portions of the Green River Formation in Wyoming. In Colorado try the Wolf Creek Pass area of the San Juan Mountains.

Zeolite Group: Stilbite

Stilbite crystals are usually short, stubby, and milky white.
PHOTO COURTESY OF THE RICE NORTHWEST MUSEUM OF ROCKS AND MINERALS

$NaCa_2Al_3Si_{13}O_{36} \cdot 16H_2O$
Family: Zeolites
Mohs: $3\frac{1}{2}$–4
Specific gravity: 2.12–2.22
Key test(s): Pearly, vitreous luster
Likely locale(s): Basalt flows and hornfels

Like heulandite, *stilbite* is the name for a series of zeolites, rather than a single species, but the difference is in the ratio of sodium and calcium present. Most stilbites are richer in calcium. Crystals are usually colorless or white, but they can be pink on occasion. Iceland has long been a prime resource for excellent stilbite specimens, but stilbite is a common zeolite and occurs alongside the rest of the zeolite family in vugs and cavities within zeolite-rich basalts.

In Idaho search Custer County, especially near Challis and the Rat's Nest Claim. In Montana try the South Fork of the Dearborn River in Lewis and Clark County, among other locales in that state. Wyoming is barren of good stilbite outside of Yellowstone, so try Colorado, with almost twenty locales in Chaffee, Jefferson, and Park Counties, among others.

Metallic Minerals

Acanthite

Acanthite specimen from the Buffalo Hump district in central Idaho
PHOTO COURTESY OF THE RICE NORTHWEST MUSEUM OF ROCKS AND MINERALS

Ag_2S
Family: Sulfides
Mohs: $2–2\frac{1}{2}$
Specific gravity: 7.2
Key test(s): Density
Likely locale(s): Rich silver districts

Acanthite is the stable version of argentite—above 360 degrees Fahrenheit, acanthite becomes argentite. Crystals are rare but attractive—they can take on a silvery elongated or tubular look. Lumps and masses are more common and are dull gray to iron black. Acanthite is very soft on the Mohs' scale, and it leaves a black streak.

Look for acanthite in Wyoming at the Yankee Jack Mine in Fremont County. In Colorado acanthite occurs in major silver-mining districts such as at Creede, Ouray, Georgetown, Leadville, Telluride, and Breckinridge. In Montana the Hecla, Philipsburg, Virginia City, and Butte districts all featured acanthite. In Idaho the Silver City, Buffalo Hump, and Yankee Fork districts all paled in comparison to the famed Silver Valley, especially the Sunshine Mine, Hecla Mine, and Bunker Hill Mine. Today, acanthite is only mined in Silver Valley.

Anglesite

Anglesite as needles and crystals from the Sunshine Mine in Kellogg, Idaho
PHOTO COURTESY OF THE RICE NORTHWEST MUSEUM OF ROCKS AND MINERALS

$PbSO_4$
Family: Sulfates
Mohs: $2\frac{1}{2}$–3
Specific gravity: 6.3
Key test(s): Density, white streak
Likely locale(s): Oxidized galena

Anglesite is sometimes confused with quartz, with small clear crystals, but anglesite shows more striations and is softer and denser. When associated with limonite, a yellow coloring may develop. When found with galena, the anglesite can darken more. When found in massive form, anglesite can be difficult to pinpoint, so check streak and density.

Anglesite is rare in Wyoming, occurring in only the Tintic and Kirwin districts. Anglesite was once common in the Berkeley Pit at Butte, and in the Jefferson County mines. Historically, Anglesite was also found throughout Colorado's silver-mining districts, and there were numerous specimens from Kellogg, Wallace, and Coeur d'Alene, Idaho. These areas are mined out in present day.

Arsenopyrite

Arsenopyrite crystals have a warped look compared to pyrite.
PHOTO COURTESY OF THE RICE NORTHWEST MUSEUM OF ROCKS AND MINERALS

FeAsS
Family: Metal sulfides
Mohs: $5\frac{1}{2}$–6
Specific gravity: 5.9–6.2
Key test(s): Density; garlic odor when crushed
Likely locale(s): Pegmatites, vein systems

Arsenopyrite is a variation on common pyrite, with an arsenic ion substituting for iron in the crystal lattice. That substitution tends to warp the crystals—where pyrite has more striking angles and lines, arsenopyrite is frequently striated and twinned. Where pyrite is a yellowy brass color, arsenopyrite is often more of a silver-white or gray color. There are at least five other variations when other ions substitute for the arsenic ion, with cobalt, nickel, stibnite, iridium, and others present in varying ratios. Arsenopyrite is not very stable and tends to oxidize, thus it is one of the culprits blamed for the reddish-orange stains draining from sulfide mines. Interestingly, this variant on fool's gold can be a key indicator to the presence of large amounts of gold, and it is associated worldwide with tin and nickel deposits.

Arsenopyrite is common in the metal deposits of the Rockies. In Wyoming the South Pass / Atlantic City gold mines featured arsenopyrite as well. Many of Montana's mining districts in Madison, Jefferson, and Park Counties contain arsenopyrite. The majority of Colorado's sulfide-metals mines contained arsenopyrite as well. Idaho contains hundreds of locales where arsenopyrite occurs.

Azurite

Velvety blue azurite with green malachite, from Leadore, Idaho
PHOTO COURTESY OF THE RICE NORTHWEST MUSEUM OF ROCKS AND MINERALS

Copper carbonate, $Cu_3(CO_3)_2(OH)_2$
Family: Metal carbonates
Mohs: $5\frac{1}{2}$–6
Specific gravity: 3.5–4.0
Key test(s): Color; hardness
Likely locale(s): Rich copper deposits

Azurite is a striking blue mineral closely associated with green malachite in copper-mining regions. The two make a pleasing combination when fresh, but azurite tends to lose its brilliance if subjected to heat, bright light, or simply too much air over too long a time. When fresh it has a bright-blue streak, which artists exploited for paint pigments during the Middle Ages in Europe. Color and streak are the two key characteristics when identifying field samples.

Idaho hosts azurite in the Coeur d'Alene district in Shoshone County. Good specimens come from the Leadore district in Lemhi County, and the Seven Devils area yields good azurite. Wyoming contains many copper-mining areas, but the Encampment district appears to offer some of the best azurite. Good azurite was actually rare from the Butte copper mines, but the nearby Troy and Hecla districts had good specimens. Azurite is widespread in Colorado, with the Leadville district noteworthy.

Bornite

Rockhounds sometimes refer to bornite as the "peacock ore" due to its many colors.
PHOTO COURTESY OF THE RICE NORTHWEST MUSEUM OF ROCKS AND MINERALS

Copper Iron sulfide, Cu_5FeS_4
Family: Metal sulfides
Mohs: 3–3¼
Specific gravity: 4.9–5.3
Key test(s): Peacock color, heft
Likely locale(s): Rich copper deposits

Rockhounds call bornite the "peacock ore" because it has so many red, purple, bronze, blue, and other hues. It is an important sulfide of copper because it is rich in copper by weight. Bornite has a metallic luster, and its streak is grayish black, but you probably won't need to streak it for identification purposes thanks to the striking play of colors. Bornite crystals are rare, as they generally appear as disseminated deposits in skarns, veins, and massive sulfide deposits.

Colorado has hundreds of bornite locales, as does Montana. Wyoming offers the usual places to look—the New Rambler, Encampment, Keystone, and Sunlight districts, especially. In Idaho search the remote mines in the Seven Devils area.

Cerussite

White cerussite crystals are heavier than quartz.
PHOTO COURTESY OF THE RICE NORTHWEST MUSEUM OF ROCKS AND MINERALS

$PbCO_3$
Family: Carbonates
Mohs: 3–3½
Specific gravity: 6.5
Key test(s): Heavy; white streak
Likely locale(s): Lead districts

Another name for cerussite is white lead ore. It is easy to distinguish from quartz, as it is heavier, and streaks white. Its rare crystals are usually small and rarely colorless. Cerussite is prone to twinning, with crystals growing out of one another. Look for cerussite whenever you find galena; the two travel together frequently. Cerussite most commonly displays as masses of bright-white material, with enough heft to distinguish it from zeolites or feldspars.

Cerussite is common in the lead-mining districts of the Rockies. Look in Chaffee, Gunnison, and Lake Counties in Colorado, especially near Leadville. The Encampment district in southeast Wyoming hosts cerussite. Custer and Lemhi Counties in Idaho both report considerable cerussite. Montana contains multiple deposits, especially in Beaverhead County and near Butte and Helena.

Chalcocite-Digenite

Digenite from the Leonard Mine in Silver Bow County, Montana
PHOTO COURTESY OF THE RICE NORTHWEST MUSEUM OF ROCKS AND MINERALS

Copper sulfide, Cu$_2$S
Family: Metal sulfides
Mohs: $2\frac{1}{2}$–3
Specific gravity: 5.5–5.8
Key test(s): Dense; black to lead-gray streak
Likely locale(s): Known copper areas

Chalcocite is another important copper ore. It rarely occurs as crystals, but when it does, it forms typically in hydrothermal vein systems, with a metallic luster and in tabular form. Chalcocite is much more common as a secondary mineral in massive oxidized zones, where the copper has leached out from other minerals. There are at least nine other copper-sulfide minerals that have varying ratios of copper and sulfur; these comprise the Chalcocite-Digenite group.

Chalcocite is common in Idaho's Hells Canyon and Seven Devils mines. Latah and Lemhi Counties also contain numerous chalcocite showings. Albany County and Carbon County in Wyoming are notable, among other locales. There are hundreds of showings in both Colorado and Montana.

Chalcopyrite

Cluster of botryoidal chalcopyrite crystals from the Silver Bow Mine in Montana
PHOTO COURTESY OF THE RICE NORTHWEST MUSEUM OF ROCKS AND MINERALS

Copper iron sulfide, $CuFeS_2$
Family: Metal sulfides
Mohs: $3\frac{1}{2}$–4
Specific gravity: 4.1–4.3
Key test(s): Crystals aren't as cubic
Likely locale(s): Known copper areas

Chalcopyrite is a common sulfide that is chemically quite similar to common pyrite, except chalcopyrite contains both copper and iron, whereas pyrite has no copper. Like pyrite, chalcopyrite is brassy and golden yellow, but chalcopyrite is softer than pyrite, and its crystal habit is a tetrahedron, while pyrite is cubic. The streak is greenish black, also similar to pyrite. There is an entire family of minerals related to chalcopyrite, for example with varying amounts of silver and gallium substituting for iron or copper, and selenium substituting for sulfur. Chalcopyrite often occurs with hydrothermal economic ore deposits that host silver and gold. It forms massive sulfide zones with other sulfides, especially pyrite, and is a primary copper ore.

Most of the metal sulfide mines and prospects in the Rockies that contain pyrite also contain chalcopyrite. There are hundreds of locales across the Rockies that supply chalcopyrite.

Cinnabar

Sugary coating of pale-red cinnabar
PHOTO COURTESY OF THE RICE NORTHWEST MUSEUM OF ROCKS AND MINERALS

Mercury sulfide, HgS
Family: Metal sulfides
Mohs: $5\frac{1}{2}$–6
Specific gravity: 3.5–4.0
Key test(s): Crimson, pink (when fresh), hard
Likely locale(s): Calderas, limestones

Cinnabar is easy to identify in the field, being a vivid red. Hexagonal crystals are quite rare; instead, you should look for reddish streaks and crusts in sulfide-rich zones. If there is enough of a sample for a streak test, cinnabar leaves a scarlet streak. In rich mercury mines with heavy concentrations of cinnabar, miners have actually noted liquid mercury deposits in small pools and puddles, so look for that. As anyone who played with a broken thermometer as a child can attest, mercury is liquid at room temperature, but the vapors are dangerous. Cinnabar is heavy and bright red, but it can be confused with ochre, rust, and even common red paint. Cinnabar samples tend to be heavier than ochre, and the streak is more vivid than hematite. When cinnabar appears agatized or opalized, rockhounds use the term "myrickite."

Idaho's Yellow Pine, Chamberlain, and Weiser districts are just some of that state's cinnabar producers. Montana is more limited, but the Boulder district in Jefferson County is noteworthy. Wyoming lists a single occurrence at the Golden Dome in Fremont County. In Colorado the Mercury Mine at Cochetopa Creek in Saguache County yields specimens.

Copper

Native copper from Butte, Montana, slowly oxidizing
PHOTO COURTESY OF THE RICE NORTHWEST MUSEUM OF ROCKS AND MINERALS

Copper, Cu
Family: Elemental metal
Mohs: $2\frac{1}{2}$–3
Specific gravity: 8.9
Key test(s): Shiny as a penny; heavy
Likely locale(s): Found with metamorphic rocks and in hydrothermal replacement deposits

Native copper, when fresh, can appear as bright as a shiny new penny. However, copper quickly tarnishes when exposed to air and starts to turn black, green, or blue. Native copper nuggets leave a characteristic copper streak and are easy to identify without much practice thanks to the copper coins in our financial system. Crystals are rare and are in the isometric habit. Copper is very malleable, meaning it bends and can be pounded, flattened, or rolled.

Idaho contains numerous native copper deposits, including good specimens from the Blackbird Mine in Lemhi County. Wyoming also has multiple copper occurrences, notably in Albany County. Colorado's Montrose County contains nice copper specimens. Montana's Butte district is particularly notable in that state.

Covellite

Covellite from the Leonard Mine near Butte in Silver Bow County, Montana
PHOTO COURTESY OF THE RICE NORTHWEST MUSEUM OF ROCKS AND MINERALS

CuS
Family: Sulfides
Mohs: $1\frac{1}{2}$–2
Specific gravity: 4.6–4.8
Key test(s): Metallic luster; lead-gray streak
Likely locale(s): Copper sulfide deposits

Because it tends to oxidize quickly, covellite is not always easy to identify. It usually forms flat, black crystals in the hexagonal system, and its blades and plates can stack up enough to resemble mica. Streak and luster differentiate it from mica, however. Covellite can also be massive, and it can occur as very small crystals. Covellite can take on a blue, iridescent color, resembling bornite, but is much softer than bornite.

Covellite is common in the copper mines of the Seven Devils area; at Buffalo Hump; and in the Coeur d'Alene District in Shoshone County in Idaho. Covellite is widespread in Montana, especially from the Butte District. In Wyoming's southeast corner, the Medicine Bow Mountains and the Ferris-Haggarty Mine in the Encampment district contain covellite. Colorado's copper districts also contain covellite.

Cuprite

Cuprite, variety chalcotrichite
PHOTO COURTESY OF THE RICE NORTHWEST MUSEUM OF ROCKS AND MINERALS

Cu_2O
Family: Oxides
Mohs: $3\frac{1}{2}$–4
Specific gravity: 6.1
Key test(s): Dark-red crystals; streak
Likely locale(s): Copper regions

Cuprite forms impressive dark-red crystals in the cubic habit, but few locales ever yield crystals large enough to facet. More common are cuprite hairs, which are fragile and unsuitable for lapidary work; even large crystals would be difficult to facet, being soft. Luster varies, from submetallic to a sharp, adamantine appearance. Penetration twins are common, and twelve-sided dodecahedrons are rare. The streak is unusual, appearing as a shiny, metallic red or metallic red brown.

There are numerous cuprite locales associated with the various copper districts of the Rockies. Albany County and Carbon County supply most of the cuprite specimens from Wyoming. Colorado, Montana, and Idaho have dozens of cuprite locales.

Enargite

Enargite with pyrite, from the Leonard Mine in Silver Bow County, Montana
PHOTO COURTESY OF THE RICE NORTHWEST MUSEUM OF ROCKS AND MINERALS

Cu_3AsS_4
Family: Sulfides
Mohs: 3
Specific gravity: 4.4
Key test(s): Distinct cleavage, metallic luster
Likely locale(s): Hydrothermal veins

Enargite typically occurs as small, tabular crystals or black, sooty masses in sulfide veins. Enargite shares similarities with several other minerals, with substitution for zinc, silver, antimony, arsenic, and others. It is often associated with quartz and pyrite and makes attractive display pieces in crystal form. Most times, however, crystals are quite small.

Enargite is common in Colorado, especially at the sulfide deposits in Gilpin, Ouray, and Rio Grande Counties. In Idaho look for enargite at Mineral in Washington County, in the Blackbird district near Cobalt, and in the Boulder Creek district in Custer County. Wyoming contains little or no enargite; Montana is rich with deposits, especially around Butte.

Galena

Massive galena from the Bunker Hill Mine in Silver Valley, Idaho (left), and large gray galena cube (right)
PHOTOS COURTESY OF THE WALLACE DISTRICT MINING MUSEUM, WALLACEMININGMUSEUM.ORG

Lead sulfide, PbS
Family: Metal sulfides
Mohs: $2\frac{1}{2}$
Specific gravity: 7.4–7.6
Key test(s): Gray, soft, stairways
Likely locale(s): Known sulfide areas

Galena is the primary ore for lead, and thus specific gravity is one of the key clues for identifying galena in the field. The dull, bluish-gray color, dull luster, and cubic, stair-stepped crystal structure are other easy clues to spot. Finally, galena is relatively soft, and you should be able to scratch it with your fingernail or a piece of calcite. Galena leaves a dark-gray streak, which is distinctive once you've seen it. It has a brittle fracture and can look like stibnite, the ore of antimony, but stibnite is harder. Cubes are common and unforgettable as well. Galena is usually found in veins and larger disseminated deposits and can mix with argentite, tetrahedrite, or sphalerite to create rich lead, silver, and zinc ores.

There are more than a thousand occurrences of galena in the Rocky Mountain states. The Kirwin, Encampment, and Sunlight districts in Wyoming are all galena producers. Colorado's Idaho Springs district has some of the hundreds of galena locales in that state. Montana also has hundreds of galena occurrences among its sulfide mines. Idaho is famous for its galena deposits in the Silver Valley.

Goethite

As goethite continues to oxidize, it takes on a yellow or orange pigment.
PHOTO COURTESY OF THE RICE NORTHWEST MUSEUM OF ROCKS AND MINERALS

FeO(OH)
Family: Iron oxides
Mohs: 5–5½
Specific gravity: 3.3–4.3
Key test(s): Crumbly; iron staining
Likely locale(s): Weathered iron mineralization

MINERALS

Goethite is the mineral residue left behind when an iron sulfide such as pyrite loses its sulfur ion. Because it is rich in iron, it serves as a valuable iron ore, and the ancient cave painters used goethite for pigments, calling it ochre. Goethite often occurs with limonite, another rusty yellow iron ore. It is typically brown, yellow, or even orange, usually as a mass but sometimes as a coating. The streak shows the same variations. Crystals are rare, but goethite easily forms pseudomorphs as it oxidizes from a sulfur-rich parent.

Goethite is scattered across the Rockies in every major metal sulfide district.

Gold

Gold foil on quartz from Colorado (left); 1.7-troy-ounce gold nugget from Murray, Idaho (right)
PHOTOS COURTESY OF THE RICE NORTHWEST MUSEUM OF ROCKS AND MINERALS

Au
Family: Elemental metal
Mohs: $2\frac{1}{2}$–3
Specific gravity: 19.3
Key test(s): Weight, color
Likely locale(s): Quartz veins

Gold is very dense, at around 19 grams per cubic centimeter, but it is also soft enough to scratch with calcite. Those two keys alone would be enough if you could find a big enough specimen to test. Instead, most prospectors see tiny specks of gold in their pans, where neither test is practical. Two minerals have earned the name "fool's gold" because they mimic gold's appearance—pyrite, which is brassy, harder, and smells like sulfur when crushed; and mica, which breaks easily with a knifepoint and tends to float. Crystals are exceedingly rare, but they are isometric with no cleavage. The density is the best clue, however. Prospectors wash gold from gravels using tools such as pans, sluices, dredges, or high-bankers when there is enough water; in the desert they use dry washers.

Gold was a major draw for miners opening up the Rockies, and nuggets remain. In 1989 placer miners recovered the 27.5-ounce Highland nugget from the Highland Mountains south of Butte. Major strikes include the Boise Basin, Florence, Pierce, and Jordan Creek in Idaho; Telluride, Ouray, Cripple Creek, and Pikes Peak in Colorado; Last Chance Gulch and Confederate Gulch near Helena, Montana; and the South Pass and Enchantment areas of Wyoming.

Hematite

Banded iron from South Pass, Wyoming (left); polished hematite "kidney stone" (right)
POLISHED HEMATITE PHOTO COURTESY OF THE RICE NORTHWEST MUSEUM OF ROCKS
AND MINERALS

Iron oxide, Fe_2O_3
Family: Oxides
Mohs: 5–6
Specific gravity: 4.9–5.3
Key test(s): Bright red on metal
Likely locale(s): Iron-rich mineralized areas

Hematite is rich in iron and is the most common iron ore thanks to dominating banded iron formations. In the field hematite can appear gray or black and has a distinct metallic luster, but after prolonged exposure to air, hematite will eventually start to show its characteristic rusty-red signature. Hematite has a striking red streak, which is one telltale sign. Crystals are varied; they can be hexagonal, tabular, or columnar. Hematite can also display a rounded, bubbly botryoidal habit. Hematite power makes up red ochre, one of the oldest color tints known to man. Similarly, yellow ochre is also hematite, but it contains extra water that results in a yellow, not red, color.

Hematite is common in Idaho, particularly in the Seven Devils district and the Sawtooth Mountains. The Blackfoot district in Latah County also contains hematite. Montana's Beaverhead County mines contain considerable hematite, as does the Boulder Batholith in Jefferson County. Wyoming's Hartville district in Goshen County and both Albany and Carbon Counties produce hematite. Colorado lists hundreds of hematite occurrences in its metal mines.

Magnetite

Magnetite in crystalline form from the Spring Mountain district of Lemhi County, Idaho
PHOTO COURTESY OF THE RICE NORTHWEST MUSEUM OF ROCKS AND MINERALS

Iron oxide, Fe_3O_4
Family: Metal oxides
Mohs: $5\frac{1}{2}$–$6\frac{1}{2}$
Specific gravity: 4.9–5.2
Key test(s): Magnetic; heavy, hard, and dark; attracts magnetic dust
Likely locale(s): Iron-rich mineralized areas

Magnetite, also known as lodestone, is easy to identify in the field thanks to its magnetic properties. It is usually dark or gray black, with a metallic luster, and leaves a black streak. Crystals are rare and usually small in the isometric habit and typically form stubby, doubly terminated octahedrons. Magnetite is quite common, as it tends to remain behind when many igneous rocks erode. Almost all rivers and streams carry some quantity of magnetite-rich "black sands" in their cracks and underneath bigger rocks. Gold panners report that there is always black sand with placer gold, but there is not always visible gold in black sands. Many prospectors save all of their black sands because, in addition to magnetite, they can usually find palladium, platinum, and other rare-earth metals in the mix. Ocean beaches typically concentrate magnetite thanks to wave action.

There are hundreds of magnetite showings in the Rockies. Good specimens of magnetite come from Idaho's Lemhi County. Montana's Georgetown district supplies decent magnetite. Both Chaffee County and Park County supply magnetite specimens from Colorado.

Malachite

Unpolished malachite showing botryoidal habit
PHOTO COURTESY OF THE RICE NORTHWEST MUSEUM OF ROCKS AND MINERALS

Copper Carbonate, $Cu_2CO_3(OH)_2$
Family: Metal silicates
Mohs: $3\frac{1}{2}$–4
Specific gravity: 4.1–4.3
Key test(s): Green; presence of azurite
Likely locale(s): Low-grade copper deposits

This mineral can be a welcome sign in otherwise perplexing or barren outcrops. Even in low concentrations, malachite leaves a telltale green stain, indicating that there is some form of mineralization present. Since it is a carbonate, it is often associated with limestones, which are calcium carbonate. In a pure form, malachite makes for a nice specimen and will take a high polish. The key characteristic for identifying malachite is its bright-green appearance. Malachite also has a light-green streak. Malachite-rich stalactites often display ornate banding, offering many lapidary possibilities.

There are hundreds of malachite locales across the Rockies. Search in known copper-producing districts such as around Butte, Montana; the Seven Devils region of Idaho; Wyoming's Encampment district; and the Leadville district in Colorado.

Meteorite–Iron

In 2007 amateur prospectors found the 8.5-ounce Cotopaxi meteorite with a metal detector in Colorado.
PHOTO COURTESY OF THE IMAGE ARCHIVES, DENVER MUSEUM OF NATURE AND SCIENCE

90 percent nickel to stony
Family: Extraterrestrial
Mohs: Varies by nickel-iron content
Specific gravity: Varies by nickel-iron content
Key test(s): Widmanstätten pattern
Likely locale(s): Anywhere, but deserts have slower oxidation

Most meteorites are actually stony, or primarily stone, but have enough iron to attract a magnet on a string, so the magnet test still works. Nickel is also a key component, measuring up to 7 percent, and many metal detectors can indicate the presence of nickel. After the magnet test, and barring a metal detector, look for thumbprint-shaped divots called regmaglypts, which are evidence of plenty of heat generated while passing through the atmosphere. Meteorites leave no streak, so a streak test that turns reddish brown is most likely hematite, and a blackish-gray streak is magnetite. Scientists can slice these rocks with a saw and use a strong acid to etch the polished surface, resulting in a characteristic X-shaped structure known as a Widmanstätten pattern.

The best place to scout for meteorites is in a known "strewn field" surrounding an observed fall. Barring that, desert environments are productive because the limited moisture ensures that the iron-rich rocks do not rust and fall apart. Many of the flat deserts in Basin and Range country offer meteorite-hunting possibilities.

Meteorite—Stony

Stony meteorite fragment from the Kelley Meteorite, a chondrite found in Logan County, Colorado in 1937
PHOTO COURTESY OF THE IMAGE ARCHIVES, DENVER MUSEUM OF NATURE AND SCIENCE

90 percent nickel to stony
Family: Extraterrestrial
Mohs: Varies by nickel-iron content
Specific gravity: Varies by nickel-iron content
Key test(s): Blackened crust
Likely locale(s): Anywhere, but deserts have slower oxidation

Most meteorites are stony, or primarily stone, but with enough iron to attract a magnet on a string. The top clues between meteorites and meteor-wrongs rely on sight. For example, look for the presence of a fusion crust, which is simply evidence that the rock burned its way through Earth's atmosphere. This will typically appear as a black skin, and if the rock has been broken or fractured, look for a bright interior beneath the black crust. Meteorites leave no streak, so a streak test that turns reddish brown is most likely hematite, and a blackish-gray streak is magnetite.

The best place to scout for meteorites is in a known "strewn field" surrounding an observed fall. Desert environments are productive because the iron-rich rocks don't rust and quickly fall apart. Many of the dry lakebeds and flat deserts in Basin and Range country offer meteorite-hunting possibilities.

Molybdenite

Rare molybdenite crystal
PHOTO COURTESY OF THE RICE NORTHWEST MUSEUM OF ROCKS AND MINERALS

Molybdenum sulfide, MoS$_2$
Family: Metal sulfides
Mohs: 1–1½
Specific gravity: 4.6–5.1
Key test(s): Green streak; soft and greasy
Likely locale(s): Interesting mineralized zones

Molybdenite is the chief ore of molybdenum. It is somewhat rare and usually occurs only as small, dull-gray lumps. It has a metallic luster that could be confused with galena, and molybdenite is dark-blue gray to lead gray. One interesting field test is that molybdenite creates a green streak when scraped on a streak plate. When big enough to conduct a scratch test, molybdenite is extremely soft, and a fingernail should scratch it easily. Crystals are rare, hexagonal, and tabular, but easily deformed. Fracture is unlikely, as it is so flexible and soft. Graphite is also lead gray but is lighter than molybdenite. Galena is also gray and soft, but it has a blue-gray streak and cubic crystal structure.

Idaho has nearly one hundred molybdenite occurrences, with Boundary County an important contributor. Albany County and Sublette County are important in Wyoming. Montana has more than one hundred molybdenite locales, with nice specimens from Beaverhead and Silver Bow Counties. Chaffee County, Lake County, and Ouray County are important in Colorado.

Platinum

Platinum cube, showing dull, silvery appearance
PHOTO COURTESY OF THE RICE NORTHWEST MUSEUM OF ROCKS AND MINERALS

Pt

Family: Elemental metal

Mohs: 4–4½

Specific gravity: 21.5

Key test(s): Silvery gray and heavy

Likely locale(s): Black sands; accessory in metals mining

Platinum is a dull white-gray metal that looks somewhat like silver but is at least twice as dense. Platinum is also far more rare than silver, occurring mostly as an accessory in black sands with other Platinum Group Metals (PGMs), which include ruthenium, rhodium, palladium, osmium, iridium, and platinum. Because platinum does not oxidize, it is useful in jewelry, and it has industrial uses as well. Most of the world's platinum comes from South Africa.

Platinum is scattered across Idaho placers, such as Rock Flat, Snake River, and Rocky Bar. Other than placers, the Stillwater Complex produces the bulk of the platinum in Montana. The La Plata, New Rambler, and Keystone districts contain the bulk of Wyoming's platinum. In Colorado the mines at Breckenridge, Telluride, Ouray, and Aspen have all contributed platinum.

Pyrite

Brassy pyrite crystals from Colorado
PHOTO COURTESY OF THE RICE NORTHWEST MUSEUM OF ROCKS AND MINERALS

Iron sulfide, FeS$_2$
Family: Metal sulfides
Mohs: 6–6½
Specific gravity: 4.9–5.1
Key test(s): Hardness
Likely locale(s): Known sulfide areas

Pyrite, or iron pyrite, is a form of fool's gold, thanks to its brassy, yellowish color. Pyrite generally forms in the cubic crystal habit and often has striations, or lines, on its crystal faces. Other times, pyrite crystals tend to twin, interlock, and form interesting masses. There are dozens of varieties of pyrite, with various replacements for the Fe ion. Pyrite tarnishes rapidly, becoming darker and somewhat iridescent as oxygen attacks the iron-sulfur bond. Pyrite leaves a brownish-black or even greenish-black streak that smells faintly of sulfur, depending on how big of a streak you make. Pyrite is hard, registering as high as 6½ on the Mohs' scale, just below quartz. Many forms of pyrite are highly collectible.

Pyrite is common in every metal-mining district where sulfides are common. Striking display pieces from Colorado are particularly interesting.

Scheelite

Orange octahedron of scheelite
PHOTO COURTESY OF THE RICE NORTHWEST MUSEUM OF ROCKS AND MINERALS

CaWO$_4$
Family: Tungstates
Mohs: $4\frac{1}{2}$–5
Specific gravity: 5.9–6.1
Key test(s): Silvery (when fresh); tarnishes quickly
Likely locale(s): Economic ore deposits

Scheelite is the most common ore for tungsten, a metal important in the steel industry. Scheelite is typically yellow to yellow orange but can be white, colorless, light gray, or other pale shades. Pure scheelite fluoresces a nice sky blue under shortwave ultraviolet light. Scheelite has a vitreous luster, has good cleavage, and forms excellent crystals that imitate diamonds. Scheelite can occur in granite pegmatites but is most common in hydrothermal veins where tin or gold is also present.

Scheelite is another metal sulfide common to the mining districts of the Rockies. In Idaho the Vulcan Mine and Tungsten Hill Mine both contain tungsten ore. Beaverhead County in Montana has several scheelite deposits, and there are many more in that state. Fremont County, Wyoming, contains several deposits, and both Boulder County and Chaffee County in Colorado contain scheelite.

Siderite

Siderite in a yellow-brown mass
PHOTO COURTESY OF THE RICE NORTHWEST MUSEUM OF ROCKS AND MINERALS

Iron carbonate, FeCO$_3$
Family: Carbonates
Mohs: $3\frac{1}{2}$–$4\frac{1}{2}$
Specific gravity: 4
Key test(s): Yellow-brown coatings
Likely locale(s): Bedded sedimentary deposits

Siderite is typically yellow, brown, or tan. Siderite crystals are rare; instead, siderite usually occurs as masses of yellow-brown masses. It occurs in hydrothermal veins and in bedded sedimentary deposits rich in calcium carbonate, but siderite also occurs as concretions in shales and sandstones. It is rich in iron by weight and is an important iron ore because it contains little sulfur. Siderite crystals are most common in pegmatites, occurring as trigonal or hexagonal tabs with vitreous, silky, or pearly luster. Siderite has a white streak.

There are numerous siderite locales across the Rockies, with Idaho and Colorado leading the way. Montana and Wyoming have limited siderite deposits. Most collectible specimens come from Colorado, such as from Eagle County.

Silver

Native silver; the pinkish tint of "horn silver" in the sliced sample, lower left

Ag
Family: Elemental metal
Mohs: $2\frac{1}{2}$–3
Specific gravity: 10.1–11.1
Key test(s): Silvery (when fresh); tarnishes quickly
Likely locale(s): Economic ore deposits

Silver is actually rare in its native state because it forms oxides such as argentite so readily. Isometric crystals are especially rare and very collectible. Silver has a characteristic shiny, metallic luster when fresh, appearing light gray or whitish gray at times. However, silver tarnishes quickly to black, brown, or yellow. Two key tests are for hardness and specific gravity. Lead forms crystals easier and has the characteristic stair-step pattern. Platinum is heavier than silver and even more rare, but some regions are noted for platinum nuggets, so do your research. Silver specimens almost always have some gold present, forming what the miners called "electrum."

South Pass, Jelm Mountain, and Laramie County all contain silver in Wyoming. Montana has many silver deposits, with the Elkhorn district supplying excellent specimens. Leadville, Colorado, supplied beautiful wire silver, some in masses. Idaho's Bunker Hill mine also supplied many native silver examples, as did Silver City in Owyhee County.

Sphalerite

Massive sphalerite from Silver Valley, Idaho (left); large sphalerite crystal on a bed of chalcopyrite and dolomite (right)
MASSIVE SPHALERITE PHOTO COURTESY OF THE WALLACE DISTRICT MINING MUSEUM, WALLACEMININGMUSEUM.ORG. CRYSTALLINE SPHALERITE PHOTO COURTESY OF THE RICE NORTHWEST MUSEUM OF ROCKS AND MINERALS

(Zn,Fe)S
Family: Metal sulfides
Mohs: $3\frac{1}{2}$–4
Specific gravity: 3.9–4.2
Key test(s): Pale-yellow or light-brown streak
Likely locale(s): Economic ore deposits

Sphalerite is the primary ore for zinc, and it is often very dark, depending on how much iron is present. It can also be green, yellowish, or red if the iron content is low. Well-formed crystals tend to be resinous or greasy black and are rare. Some varieties fluoresce.

Sphalerite makes up much of the zinc ore in Idaho's Silver Valley. Wyoming's Park County hosts multiple sphalerite deposits. Montana and Colorado both contain hundreds of sphalerite locales in the usual sulfide-rich mining districts for those states.

Stibnite

Massive antimony ore from Silver Valley, Idaho (left); attractive blades of stibnite in more crystalline form (right)

MASSIVE ANTIMONY ORE PHOTO COURTESY OF THE WALLACE DISTRICT MINING MUSEUM, WALLACEMININGMUSEUM.ORG; STIBNITE BLADES PHOTO COURTESY OF THE RICE NORTHWEST MUSEUM OF ROCKS AND MINERALS

Antimony sulfide, Sb_2S_3

Family: Metal sulfides

Mohs: 2

Specific gravity: 4.6

Key test(s): Soft, slender, gray crystals

Likely locale(s): Hydrothermal deposits

Stibnite is the principal ore of antimony, and it is highly toxic. It has a metallic-gray luster and streak and is exceedingly soft, so it is easy to crush into a powder. It was known as "kohl" in the ancient Middle East and served as a cosmetic eyeliner, which would not have been healthy. Stibnite forms long, slender crystals, some quite stunning, and it makes an attractive addition to your collection. It is often associated with arsenic minerals such as orpiment, realgar, and arsenopyrite in metal-rich hydrothermal vein systems.

Large deposits of stibnite are rare, but small deposits are scattered across the Rockies. Colorado hosts a few dozen locales in Mineral County and elsewhere. The antimony mines on Stibnite Hill in Montana's Sanders County produce good specimens. Stibnite is rare in Wyoming, but Idaho has dozens of prospects and mines, especially including the appropriately named town of Stibnite.

Tetrahedrite

Massive tetrahedrite from Silver Valley, Idaho (left); attractive blades of tetrahedrite from Peru (right).
MASSIVE TETRAHEDRITE PHOTO COURTESY OF THE WALLACE DISTRICT MINING MUSEUM,
WALLACEMININGMUSEUM.ORG; CRYSTALLINE TETRAHEDRITE PHOTO COURTESY OF THE RICE
NORTHWEST MUSEUM OF ROCKS AND MINERALS.

$(Cu,Fe)_{12}Sb_4S_{13}$
Family: Metal sulfides
Mohs: $3\frac{1}{2}-4$
Specific gravity: 4.97
Key test(s): Distinctive crystals
Likely locale(s): Base metal zones

This is another mineral that represents a series, being the anti-mony end of a changing formula with the arsenic mineral tennan-tite at the other end of the scale. Another variation, freibergite, can be quite rich in silver. In its pure form, tetrahedrite forms unmistakable tetrahedrite crystals, hence the name, with distinct triangles sometimes flattened at each point. Tetrahedrite com-monly occurs in massive form, being dark gray before rusting and turning reddish or yellowish.

Idaho hosts dozens of tetrahedrite locales, including the famed Silver Valley where it was a major ore. Montana also contains doz-ens of good locales, with good specimens from Silver Bow County and the Philipsburg district. Wyoming is limited for tetrahedrite, with most samples in Park County and Albany County. Colorado is loaded with tetrahedrite sources, with excellent crystals coming from the famed Sweet Home Mine in the Alma district.

Vanadinite

Vanadinite from Montana
PHOTO COURTESY OF THE RICE NORTHWEST MUSEUM OF ROCKS AND MINERALS

$Pb_5(VO_4)_3Cl$
Family: Apatites
Mohs: 3
Specific gravity: 6.7–7.1
Key test(s): Color, heft
Likely locale(s): Oxidized zones with galena

Vanadinite is usually a reddish-to-orange mineral, usually found as small, crystalline masses or coatings. Crystals are hexagonal, but large crystals are rare. Its streak is white to yellow, and its luster is resinous. The presence of lead gives vanadinite a heavy nature, helping distinguish it from mimetite and pyromorphite, but the key distinguishing characteristic is the color. Vanadinite primarily forms in the oxidized zones of lead deposits.

Vanadinite is a rare ore in the Rockies. Colorado hosts a handful of deposits, in Leadville and Breckinridge among others. Beaverhead and Broadwater Counties in Montana contain a dozen or so locales, while Wyoming is apparently empty. Lemhi County in Idaho contains scattered vanadinite deposits.

Veszelyite

Attractive blue veszelyite crystals from the Black Pine Mine in the Philipsburg district of Montana
PHOTO COURTESY OF THE RICE NORTHWEST MUSEUM OF ROCKS AND MINERALS

$(Cu,Zn)_2Zn(PO_4)_2 \cdot 2H_2O$
Family: Phosphates
Mohs: $3\frac{1}{2}$–4
Specific gravity: 3.42
Key test(s): Color
Likely locale(s): Oxidized zones of base metal deposits

Veselyite is a rare secondary mineral of zinc and lead deposits, and its attractive blue color makes it a prized specimen for collectors. Crystals are anywhere from green to dark blue, with the darker colors highly sought. It has a vitreous luster, green to white streak, and attractive monoclinic crystals.

Montana contains a single veszelyite deposit in the Philipsburg district, at the Black Pine Mine in the John Long Mountains of Granite County.

Vivianite

Dark-gray vivianite from the Blackbird Mine near Cobalt in Lemhi County, Idaho
PHOTO COURTESY OF THE RICE NORTHWEST MUSEUM OF ROCKS AND MINERALS

$Fe_2 + Fe_{22} + (PO_4)_2 \cdot 8H_2O$
Family: Phosphates
Mohs: $1\frac{1}{2}-2$
Specific gravity: 2.68
Key test(s): Soft; darkly oxidized
Likely locale(s): Oxidation zones; pegmatites

Vivianite is typically brown or black, usually in massive form or even in concretions. Crystals are prized, and while fresh specimens are pale green or even clear, this mineral quickly begins oxidizing and eventually turns black. Vivianite is quite soft, so try scratching it with a fingernail or penny if you don't have any gypsum handy.

Montana's Berkeley Pit in Butte yielded vivianite. Mica Mountain in Idaho's Lemhi County produced some specimens, as did the Blackbird district near Cobalt, Idaho. Silver City, Idaho, also produced some vivianite specimens. Wyoming is barren, but Colorado's Leadville district and several others in that state still produce striking examples.

IV
GEMS

Ammolite

Raw ammolite (left) and finished jewelry (right)
RAW AMMOLITE PHOTO FROM THE DISPLAY OF THE GEMOLOGICAL INSTITUTE OF AMERICA

CaCO$_3$
Family: Carbonates
Mohs: $4\frac{1}{2}$–$5\frac{1}{2}$
Specific gravity: Variable
Key test(s): Iridescence
Likely locale(s): Ammonite fossil locales

Ammolite is a vivid, multicolored variety of ammonite fossil that has not only preserved but enhanced the nacre, or colored shell. If you've ever seen an abalone shell, for example, you know the beautiful colors that display. Ammolite, fossilized ammonite shell, is highly prized by jewelers for the iridescent play of colors possible. Striking orange, green, red, yellow, and blue colors all vie for attention in ammolite. Since ammolite is often very thin and quite fragile, jewelers and lapidarists layer it with harder material for protection.

Most of the world's ammolite supply originates in Canada. However, equivalent rocks to the Cretaceous Bearpaw Formation of Canada extend into Colorado.

GEMS

Aquamarine

Aquamarine vug called Diane's Pocket, discovered in 2004 by prospector Steve Brancato at Mount Antero, Colorado
PHOTO COURTESY OF THE IMAGE ARCHIVES, DENVER MUSEUM OF NATURE AND SCIENCE

$Be_3Al_2(SiO_3)_6$
Family: Beryl
Mohs: $7\frac{1}{2}$–8
Specific gravity: 2.8
Key test(s): Hardness
Likely locale(s): Granite pegmatites

Aquamarine is another gem variety of beryl, like emerald, heliodor, morganite, red beryl, and goshenite. Aquamarine is blue, and when deep, striking blue, it is highly sought as a gemstone. It is quite rare, thus worth prospecting for. The hardness of around 8 makes it hard to take a streak, but it is white. The luster is vitreous to resinous, cleavage is poor, and it does not fluoresce. It is usually associated with granite pegmatites.

Colorado's Mount Antero is the most famous aquamarine producer in the United States. Check out *Rockhounding Colorado* (FalconGuides, 2004) for more information.

Diamond

Loose diamonds showing tetrahedral shape
PHOTO COURTESY OF THE RICE NORTHWEST MUSEUM OF ROCKS AND MINERALS

Carbon, C
Family: Native mineral
Mohs: 10
Specific gravity: 3.5
Key test(s): Hardness
Likely locale(s): Associated strictly with kimberlite pipes

Diamond specimens are clear, yellow, pink, blue, purple, brown, and even black. The hardness test is the best indicator—pure, strongly crystallized diamond, at 10 on the Mohs' scale, scratches everything. Another good indicator is that diamond has an adamantine, or outstanding, luster. The crystal habit is octahedral, and cleavage is perfect in four directions, which is rare. Clear agate, and clear gem-quality quartz such as Herkimer "diamonds," look similar to actual diamonds but are hexagonal and will not cleave.

Diamonds are associated with kimberlite pipes along the Wyoming-Colorado border, and there has been considerable exploration, staking, and for a limited time, production. However, the Kelsey Lake Diamond Mine closed in 2002. In Idaho miners discovered three small diamonds in the placers at Little Goose Creek near New Meadows in Adams County. In *Gemstones of North America* (1959), John Sinkankas reports minor diamond finds in the gold placer mines of Montana at Nelson Hill in Deer Lodge County, Greenhorn Gulch, Madison County, and Grasshopper Creek in Beaverhead County.

Star Garnet

Polished star garnets from Emerald Creek, Idaho
PHOTO COURTESY OF JENNIFER DICKISON

$X_3Y_2(SiO_4)_3$ Almandine
Family: Metal silicates
Mohs: $6\frac{1}{2}$–$7\frac{1}{2}$
Specific gravity: 3.6–4.3
Key test(s): Hackly fracture, hardness
Likely locale(s): Schist; black sands

Gem-quality garnet is somewhat rare, and adding further value is a tendency for impurities to line up and create four-sided or six-sided stars. White rutile is the source of the star asterism at Emerald Creek. The US Forest Service operates a fee-dig operation at Emerald Creek, Idaho, where the public can wash the gravels for these prizes. Polishing the garnets to reveal the prized stars requires patience, experience, and good equipment.

Several areas around Clarkia, Idaho, produce star garnet rough—check out *Rockhounding Idaho* (FalconGuides, 2010) for more information. Most other garnet locales in the Rocky Mountain states do not yield gem-quality material unless found in pegmatites.

Jade

Jade specimens from Wyoming's jade fields

Manganese silicate, MnSiO$_3$
Family: Silicates
Mohs: $6\frac{1}{2}$–7 for jadeite; $5\frac{1}{2}$–6 for nephrite
Specific gravity: 2.9–3.1
Key test(s): Can't be scratched by a knife; botryoidal
Likely locale(s): Mafic rocks

There are two main varieties of jade. Both are amazingly strong due to interlocking crystals that form nearly unbreakable bonds. First, there is classic jadeite, a sodium-rich, aluminum-rich pyroxene, found mostly in Burma and favored by Chinese rulers for centuries. Jadeite is not quite as hard as quartz. Second, there is nephrite jade, a type of amphibole, also famous in China but more commonly associated with North American deposits in Wyoming, British Columbia, Washington State, and California. Nephrite is slightly softer than jadeite and usually occurs only as white or shades of green. A high-quality steel knife blade cannot scratch either jadeite or nephrite. Jadeite, which can be purple, blue, lavender, pink, or vivid green, is the more highly prized, but that is not a fixed rule, as rare, "mutton-fat" white nephrite jade commands fabulous prices.

Wyoming's jade fields occupy a zone in the central part of the state near Jeffrey City, and there are multiple collecting locales described in *Rockhounding Wyoming* (Globe Pequot Press, 1996).

GEMS

Precious Opal

Spencer, Idaho, boasts some of the finest precious opal in the world.
PHOTO COURTESY OF THE RICE NORTHWEST MUSEUM OF ROCKS AND MINERALS

SiO$_2$•nH$_2$O
Family: Metal silicates
Mohs: $5\frac{1}{2}$–$6\frac{1}{2}$
Specific gravity: 2.0–2.2
Key test(s): Play of color; luster
Likely locale(s): Tuff deposits

Precious opal is a rare form of common opal that owes its stunning color to the way its silica spheres are stacked and packed, diffracting the light and causing the color interplay. Precious opal is relatively soft, has no crystal structure and no cleavage, and usually occurs in veins and nodules. The streak is white. Precious opal comes in a variety of forms: standard precious opal, displaying all colors of the rainbow, usually after replacing wood; fire opal, usually red or orange, derived from thin seams between lava flows; black opal, very dark in color but with a nice play of colors; and white opal, which is usually white but also exhibits a full play of colors.

The Spencer Opal Mine, a fee-dig operation located near I-15 by Spencer, Idaho, is by far the best place to go for world-class precious opal. Cedar Rim in central Wyoming also hosts abundant opal deposits, some of which Wyoming geologist W. Dan Hausel reported to be either fire opal or gem-quality precious opal.

Sapphire

Beautiful blue, untreated Yogo sapphires from Montana's famed Yogo Gulch, rough and polished
PHOTO COURTESY OF THE GEM GALLERY, BOZEMAN, MONTANA, GEMGALLERY.COM

Aluminum oxide, Al_2O_3
Family: Oxides
Mohs: 9
Specific gravity: 3.9–4.0
Key test(s): Hardness
Likely locale(s): Pegmatites, dikes, and metamorphic zones

Sapphire and ruby are both varieties of gem-quality corundum, but unlike rubies, which are typically red, a sapphire is usually some form of blue, although sapphires can be clear, light gray, or even dark gray. The blue color comes from iron and titanium impurities that affect ion charges, color absorption . . . it all gets very technical. The inclusion of rutile needles results in a "star" effect for properly polished sapphires and is highly desirable. Note that sapphires are one of the easiest gems to treat with heat and enhance or change the color. Sapphires are typically associated with pegmatites and can be difficult to separate from host rock.

Gold miners found rubies at Rocky Flat, near Meadows, Idaho; that locale also yielded sapphire. Montana has historically been the big sapphire producer in North America, with Yogo Gulch the most coveted. Spokane Bar, near Helena, has also produced major alluvial sapphire. Check out *Rockhounding Montana* (FalconGuides, 2006) for more information.

Topaz

Large rough topaz will scratch quartz.
PHOTO COURTESY OF THE RICE NORTHWEST MUSEUM OF ROCKS AND MINERALS

Aluminum silicate, $Al_2SiO_4(F,OH)_2$
Family: Silicates
Mohs: 8
Specific gravity: 3.4–3.6
Key test(s): Pink (when fresh), hard
Likely locale(s): Colorado

Topaz is an interesting mineral that does not seem to get much respect for its natural state. Gem traders sometimes irradiate topaz to produce results that are more pleasing. Typically, topaz is clear when pure, but impurities can cause it to appear white, light gray, or even pink or yellow. Laboratory treatments result in more stunning blues and orange colors. Topaz has a glassy to vitreous luster, and forms stubby, prismatic crystals in the orthorhombic crystal system, with excellent terminations. The best test in the field is its hardness—it will scratch quartz. Cleavage is excellent in one direction, and striations are common lengthwise on crystal faces, but topaz often occurs in massive lumps. Topaz forms at high temperatures, among silica-rich igneous rocks such as rhyolite, or in cavities within granite pegmatites.

Topaz occurs in Colorado in the Front Range region, in the Tarrayall region, at Lake George, and elsewhere. In Idaho rockhounds have found topaz at Paddy Flat and Dismal Swamp. Southwest Montana reports topaz in volcanic fields. Sinkankas mentions topaz at the headwaters of the Bighorn River.

Turquoise

Polished turquoise can be creamy blue or contain black veins and inclusions
PHOTO COURTESY OF THE RICE NORTHWEST MUSEUM OF ROCKS AND MINERALS

$CuAl_6(PO_4)_4(OH)_8 \cdot 4\text{-}5H_2O$
Family: Phosphates
Mohs: 5–7
Specific gravity: 3.5–4.0
Key test(s): Blue; waxy luster
Likely locale(s): Veins, seam fillings; near copper mines

Natural turquoise rarely forms crystals, occurring instead as nuggets and filled-in fractures in the host rock. Turquoise frequently forms veins and nodules as a secondary replacement mineral. Turquoise has a waxy luster, a faint, bluish-white streak, and the powder is soluble in hydrochloric acid. Under long-wave UV light, turquoise may fluoresce. Black limonite veining is common. Some of the minerals confused with turquoise include chrysocolla, which is much softer, and variscite, which is usually greener and softer. Look for turquoise in known copper-producing areas where phosphates are also common.

Other than a minor showing in the Butte district of Silver Bow County in Montana, Colorado is the only state in the Rockies to host turquoise. Deposits of note include the King Turquoise Mine in Conejos County, the Turquoise Chief Mine in Lake County, and Villa Grove Turquoise Mine in Saguache County. Teller County also hosts numerous showings.

Glossary

alluvium: Dirt, usually. Stream and river deposits of sand, mud, rock, and other material. Sometimes sorted, if laid down in deep water; otherwise can be unsorted if deposited during floods, earthquakes, etc. If glaciers were involved, the term "till" is used.

anthracite: The hardest and most intensely metamorphosed form of coal.

arkose: Sandstone that has many unsorted, broken-up pieces of feldspar and quartz. Usually hard and not easily eroded.

basement: The "lowest" and oldest rocks around, usually metamorphic and frequently dating to the Precambrian or Paleozoic age. They are usually less prone to erosion and make up mountain ranges and stunning cliffs. "Basement" refers to their placement at the bottom of a stratigraphic table.

batholith: General term that refers to extremely large masses of coarse intrusive rock such as granite. Any rock formation over 100 km^2 (39 square miles) is considered a batholith.

bedding: The tendency of sedimentary rocks such as sandstone to reside in visible zones or marker beds, similar to tree rings.

bleb: A round or oval cavity, air bubble, hole, or vesicle, usually in basalt, and sometimes filled with opal, agate, or chalcedony.

chemical sediment: Refers to the way certain limestones and dolomites precipitate material such as calcium carbonate, which falls to the bottom of the sea or bay and accumulates.

clasts: Catchall term for the clay, silt, sand, gravel, cobbles, and boulders that make up nonchemical sedimentary rocks. The size of the clasts then determines the name of the rock.

clay: Usually refers to the smallest mineral fragments, smaller than 2μm or $\frac{1}{255}$ millimeters.

cobble: Fancy term for rocks between a pebble and a boulder. The exact definition of a cobble is anything from 64 to 256 millimeters in size.

contact metamorphism: The result of a hot igneous intrusion on the country rock. The contact zones between the intrusion and the surrounding rock can sometimes house interesting mineralization.

density: Defines the weight per an agreed unit of volume. By weighing the sample and then dunking it in water and measuring the volume of water displaced, we get the density measured in grams per cubic centimeter. The general term "heft" refers to a field test for how dense a typical-size hand specimen feels.

diatomite: Usually a white, chalky deposit that, upon microscopic inspection, turns out to be composed of tiny diatom fossils. These beds can often host common opal, precious opal, and zeolite deposits.

drift: General term for glacial deposits composed of jumbled debris. Outwash plains and terraces are usually sorted, whereas tills and moraines are unsorted.

dry wash: The sign of a seasonal stream that dries up during the summer months. These can be interesting for rockhounds, as specimen sizes are usually bigger because they have not been severely eroded.

eolian: General term for wind deposits such as loess, sand sheets, ripples, and dunes, but also can refer to wind processes such as dust storms, sandblasting, and desert varnish.

eon: The longest division of geologic time is the super-eon. The model is: super-eon —> eon —> era —> period —> epoch —> age. Thus, we are in the Holocene epoch of the Quaternary period of the Cenozoic age of the Phanerozoic eon of the Cambrian super-eon.

epoch: Shorter subdivision of a geologic period, usually corresponding to observed stratigraphy in the field.

era: The four main geologic eras, from oldest to youngest, are the Precambrian, Paleozoic (which starts with the Cambrian), Mesozoic (the age of dinosaurs), and Cenozoic (ours).

erosion: The forces and processes that continually grind down mountains and move their debris downwind or downhill.

evaporite: As bodies of water dry up under desert conditions, they frequently get white or light-brown crystalline rings around the edges. These minerals are usually pure salt (halite, or sodium chloride) or a related halide, plus borax or gypsum also.

exfoliation: This term refers to the way rocks, and in particular granite, tend to slough off skins or layers of outer rock, like an onion. The result is usually a rough, rounded shape, rather than angles and edges.

facies: Field term used to describe how sedimentary rocks can be identified by the way they were deposited. The term "biofacies" describes the distinct fossil assemblage, while the term "lithofacies" could describe the similarities in a rock's clast size. There could be several distinct facies identifiable in the field that make up an overall formation.

float: Describes the difference between rock samples hammered from an outcrop, and thus with a known origin, and samples that exist as cobbles or boulders and not attached to bedrock. Prospectors are able to trace float to its source outcrop.

flood basalt: Refers to the way basalt tends to pour out of cracks and vents and form rivers of liquid rock, and thus create plateaus of flat, layered deposits. By comparison, andesites pile up.

flow cleavage: Describes the tendency of metamorphic rocks to arrange flat, elongated crystals into a parallel structure.

formation: This is a key term to understand in field geology. Geologists assign formation names to mappable, recognizable rock assemblages and also note a "type" locale that defines the rest of the unit. Formations can be lumped together into groups or even supergroups. To become a formation, a group of similar rocks must be big enough to be worth the bother, must be the same age, must share some key similarity, and must be traceable across the surface. Some formations are divided into members.

geologic cycle: The continuous cycle of destruction, recycling, and rebirth that defines the way the earth's crust works. There are countless variations on the scenario, but in general, rocks are created, such as by a volcano, eroded, built back into sedimentary deposits, subducted, cooked, melted, then turned back into lava, and erupted again.

graded bedding: Sedimentary rock term to describe the way a creek or river deposits alluvium, pebbles, and boulders in a typical sequence, from coarse to fine. Look for coarse conglomerates at the bottom and fine siltstones at the top.

gravel: The best place for pebble pups and rockhounds to search for interesting material. Gravels usually consist of pebbles, cobbles, and boulders, in various ratios, and also contain varying amounts of sand and silt. Gravel bars refresh with each season and provide clues to the surrounding geology.

hydrothermal vein: Best spot to investigate for interesting minerals. These hot, chemical-rich solutions, usually quartz, can either find an existing crack in country rock or create their own. If they cool slowly enough, hydrothermal veins can create ore deposits with large crystals that are prized by collectors.

ignimbrite: The igneous rock created when hot volcanic ash and breccias pour out of a volcano or vent and are too heavy to drift away as an ash cloud.

intrusion: Catchall term for the various granites, diorites, and related rocks that bulldoze their way through the earth's crust but never reach the surface. Cooling in place quickly results in fine-grained material; cooling slowly gives the individual elements more time to build up into larger crystals.

laccolith: This is a small intrusion that squeezes in horizontally between beds and builds itself out laterally. These deposits usually have a neck, where material fed in, and a dome, depending on how forceful the intrusion was.

lahar: When volcanic eruptions mix with melted glaciers, lakes, and snow, the result is a dangerous mudflow called a lahar. The material can flow quickly, but it will soon solidify into a concrete-hard mass of jumbled-up ash, fragments, and pumice.

lava: Catchall term for extruded molten rock that includes basalt, andesite, rhyolite, dacite, and others.

lode: The prized zone of rich, extended mineralization that usually ensures a successful mining operation. The term is reserved for larger vein networks that cover significant ground.

luster: Term for the visual appearance of a mineral's lighted surface. The way minerals reflect light can be helpful for identification, but terms such as "metallic" and "waxy" are not completely standardized.

mafic mineral: These are the dark, heavy minerals that are rich in iron and magnesium, such as pyroxenes, amphiboles, and olivines.

magma: The molten lava that eventually forms igneous rock when it cools. Magma that cools without eruption is called an intrusion; otherwise it is extrusive.

magma chamber: The source cavity or reservoir for magma traveling up through the earth's crust.

mass spectrometer: The one instrument you wish you had, because it can count ions and provide their exact distribution. An inexpensive, handheld mass spectrometer would revolutionize field geology.

Mohs' scale of hardness: The observation-based method of ranking a mineral by what it can scratch and what, in turn, can scratch it. Thus, diamond is alone at the top of the list at 10, and it can scratch corundum, which can scratch topaz, which can scratch quartz, at 7. The remaining minerals are: orthoclase feldspar, 6; apatite, 5; fluorite, 4; calcite, 3; gypsum, 2, and talc, the softest, at 1.

native metal: Metals in their purest form are not significantly combined with oxides, sulfides, carbonates, or silicates, and are thus native. Gold, silver, copper, platinum, and mercury are examples.

oil shale: A dark, organic-rich shale that sometimes contains enough petroleum-based ingredients to burn.

oolite: Refers to the small, round form taken when calcium carbonates start to coat sand grains and roll around in a lime-rich sea bottom.

ore: The term used to describe a viable mineral deposit that is worth mining. Usually refers to a metal-based mineral that must be milled.

Original Horizontality: This term refers to the idea proposed by Nicholas Steno (1638–86) that sedimentary rocks are laid down flat. Since many sandstone beds are currently tilted, this simple concept meant other forces were at work.

outcrop: A cliff, ledge, or other visible clue to the rock formations below. Exposed, mappable basement rock.

pegmatite: A key igneous rock, usually found as a vein or dike, with very large grains of mica, feldspar, and tourmaline, among others. Pegmatites tend to form cavities and vugs where large crystals can accumulate without crowding into one another and damaging their crystal structure. Pegmatites sometimes host smoky quartz, beryl, topaz, aquamarine, and other exotic gems.

pelagic sediment: Catchall term for the fine sediments that slowly accumulate in deep marine environments. Rather than relying on silt or clastic debris, this material is predominantly derived from shells of microscopic organizations such as foraminifera. Accumulation rates are as slow as 0.1 centimeter per 1,000 years.

reaction series: Refers to the behavior of a cooling magma where some minerals form at high temperatures, and others as temperatures cool. These conditions are observable in a laboratory setting.

regolith: Another catchall term, this time describing the various rock fragments and erosional debris that lie on bedrock, including alluvium, clastics, and others.

replacement deposit: Describes a particular type of ore deposit where hot, circulating solutions first dissolve a mineral to form a cavity, then fill the void with a new material.

sedimentary structure: Describes the various relicts of a sedimentary rock's deposition, such as ripples, cracks, bedding, zones, layers, etc.

stratification: The tendency of sedimentary rocks to form in flat, parallel sequences that are mappable at the surface for considerable distances. Geologists identify patterns and unravel the sequence of events represented in the strata.

stratigraphic column: Stratigraphy is the science and study of sedimentary rock outcrops. Stratigraphers draw stratigraphic columns that pictorially represent the measured or inferred relations of rock outcrops, especially age. Metamorphic and igneous rocks occasionally show up at the bottom in stratigraphic columns, as it is primarily a tool to understand sedimentary rocks.

streak: Refers to the color of the powdered mineral dust left behind when a mineral is scraped across a streak plate. The color of this fine rock powder is a more true reflection of the mineral than visual appearance. Since a streak plate is about 7 on the Mohs' hardness scale, only minerals less than 7 can easily be tested by streak.

Superposition, Principle of: Another theory traced to Nicolas Steno (1638–86), who pointed out that a rock formation or strata that sits on top of another distinct layer must usually be younger than the rock below. The older rock will always be at the bottom, except in rare conditions.

talus: The impressive accumulation of debris below a cliff or prominent outcrop. Because the rocks are shifting constantly and continuing to accumulate, few plants can get a foothold. Also called scree.

tectonics: Geologic theories of how the earth's crust continually moves and shapes new rocks. To compare planets, on Mars there is little evidence of tectonics, and there is basically one volcano, Olympus Mons, rising 13 miles (21 km) above the planet surface. On Earth the plates continually shift, resulting in arcs, troughs, and collision zones.

texture: Describes a rock's grain size, crystal size, whether the grains are uniform or variable, whether the grains are rounded or angular, and whether there is any evidence of orientation to the grains.

till: Jumbled mess of glacial debris, with little to no recognizable bedding present and sediment sizes ranging from rock flour and clay all the way to massive boulders.

tuff: The term used to describe ash, pumice, volcanic breccias, and other debris. Some ash beds yield petrified wood and fossil leaves; others are welded solid.

ultramafic rock: Igneous rocks such as dunite, peridotite, amphibolite, and pyroxenite that consist of primarily mafic minerals and have less than 10 percent feldspar.

volcanic ash: The fine rock fragments and glassy, angular material ejected from a volcano into the air.

volcanic breccia: A pyroclastic rock made up of angular fragments that show little or no sign of waterborne movement. Particle sizes are greater than 2 millimeters in diameter.

xenolith: Literally, "foreign rock" where a piece of country rock is picked off the walls or otherwise incorporated into a rising or spreading dike or intrusion.

zeolite: Common aluminosilicates formed in volcanic rocks, such as basalt, where alkaline groundwaters circulate at low temperatures and create a ringed "molecular sieve" structure.

Index

A

acanthite, 101
actinolite, 52
Ada County (Idaho), 86
Adams County (Idaho), 137
agate
  cream, 53
  dendritic (moss), 54
  Dryhead, 55
Albany County (Wyoming), 17, 64, 68, 76, 96, 107, 110, 112, 117, 122, 130
Alden, Andrew, xv
Alder, Montana, 95
algae, fossil, 43
Alma, Colorado, 90, 130
Alma Rose, 90
amethyst, 56
ammolite, 135
amphibolite, 24
Anderson pegmatite, 93
Andes Mountains, 3
andesite, 3
anglesite, 102
Apache tears, 7
apatite, 57
aquamarine, 136
aragonite, 58
argillite, 23
arsenopyrite, 103
Artist's Point, xi
Aspen, Colorado, 123
Atlantic City, Wyoming, 103
augite, 59
azurite, 104

B

Bald Mountain, Montana, 89
barite, 60
basalt, 2
Basin and Range Province, 7, 120, 121

Bathtub Mountain, 94
Bay Horse district, 64
Bearpaw Formation, 36, 135
Beaverhead County (Montana), 95, 106, 117, 122, 125, 131, 137
Belt Supergroup, 43
Benewah County (Idaho), 94
Berkeley Pit, 102, 133
beryl, 61
Big Horn County (Wyoming), 57
Big Lost River, 56
Big Southern Butte, 7–8
Bighorn Formation, 34
Bighorn River, 142
birds, fossil, 47
Blackbird district, 113
Blackbird Mine, 110, 133
Blackfoot district, 117
Black Pine district, 87
Black Pine Mine, 132
Blue Forest, 46
Blue Mountains, 29
Boise Basin, 116
Boise County (Idaho), 88
bornite, 105
Boulder Batholith, 11, 61, 68, 96, 117
Boulder County (Colorado), 71, 72, 125
Boulder Creek Batholith, 11
Boulder Creek district, 113
Boulder district, 109
Boulder Mountains, 11
Boundary County (Idaho), 122
Brancato, Steve, 136
breccia, 38
Breckinridge, Colorado, 101, 123, 131
Broadwater County (Montana), 131
Bruneau Canyon, 74
Bruneau, Idaho, 79
Buffalo Hump district, 101, 111
bugs, fossil, 47
Bunker Hill Mine, 101, 114, 127

About the Author

Garret Romaine, an avid rockhound, fossil collector, and gem hunter, is the author of *Rockhounding Idaho*, *Gem Trails of Oregon*, and *Gem Trails of Washington*. He was a columnist for *Gold Prospectors* magazine for fifteen years, and he has created numerous instructional and informational videos for YouTube in his series "Garret's World of Geology." Garret is a member of the Board of Directors of the Rice Northwest Museum of Rocks and Minerals in Hillsboro, Oregon.